Peter Kaiser

Fabrication and Characterization of Extracellular Matrix Nanofibrils

Peter Kaiser

Fabrication and Characterization of Extracellular Matrix Nanofibrils

Fibronectin, Collagen and Laminin

Südwestdeutscher Verlag für Hochschulschriften

Impressum/Imprint (nur für Deutschland/ only for Germany)
Bibliografische Information der Deutschen Nationalbibliothek: Die Deutsche Nationalbibliothek verzeichnet diese Publikation in der Deutschen Nationalbibliografie; detaillierte bibliografische Daten sind im Internet über http://dnb.d-nb.de abrufbar.

Alle in diesem Buch genannten Marken und Produktnamen unterliegen warenzeichen-, markenoder patentrechtlichem Schutz bzw. sind Warenzeichen oder eingetragene Warenzeichen der jeweiligen Inhaber. Die Wiedergabe von Marken, Produktnamen, Gebrauchsnamen, Handelsnamen, Warenbezeichnungen u.s.w. in diesem Werk berechtigt auch ohne besondere Kennzeichnung nicht zu der Annahme, dass solche Namen im Sinne der Warenzeichen- und Markenschutzgesetzgebung als frei zu betrachten wären und daher von jedermann benutzt werden dürften.

Verlag: Südwestdeutscher Verlag für Hochschulschriften Aktiengesellschaft & Co. KG
Dudweiler Landstr. 99, 66123 Saarbrücken, Deutschland
Telefon +49 681 37 20 271-1, Telefax +49 681 37 20 271-0
Email: info@svh-verlag.de
Zugl.: Heidelberg, Universität, Diss., 2009

Herstellung in Deutschland:
Schaltungsdienst Lange o.H.G., Berlin
Books on Demand GmbH, Norderstedt
Reha GmbH, Saarbrücken
Amazon Distribution GmbH, Leipzig
ISBN: 978-3-8381-1419-4

Imprint (only for USA, GB)
Bibliographic information published by the Deutsche Nationalbibliothek: The Deutsche Nationalbibliothek lists this publication in the Deutsche Nationalbibliografie; detailed bibliographic data are available in the Internet at http://dnb.d-nb.de.

Any brand names and product names mentioned in this book are subject to trademark, brand or patent protection and are trademarks or registered trademarks of their respective holders. The use of brand names, product names, common names, trade names, product descriptions etc. even without a particular marking in this works is in no way to be construed to mean that such names may be regarded as unrestricted in respect of trademark and brand protection legislation and could thus be used by anyone.

Publisher: Südwestdeutscher Verlag für Hochschulschriften Aktiengesellschaft & Co. KG
Dudweiler Landstr. 99, 66123 Saarbrücken, Germany
Phone +49 681 37 20 271-1, Fax +49 681 37 20 271-0
Email: info@svh-verlag.de

Printed in the U.S.A.
Printed in the U.K. by (see last page)
ISBN: 978-3-8381-1419-4

Copyright © 2010 by the author and Südwestdeutscher Verlag für Hochschulschriften Aktiengesellschaft & Co. KG and licensors
All rights reserved. Saarbrücken 2010

Contents

Abstract	1
I Introduction	**3**
1 Mechanosensing in the Extra Cellular Matrix	**5**
2 Extra Cellular Matrix (ECM) Proteins	**8**
2.1 Fibronectin (FN)	8
2.1.1 Structure	8
2.1.2 Function	10
2.1.3 Fibrillogenesis *in vitro*	11
2.1.4 Mechanical Properties	12
2.2 Collagen (COL)	13
2.2.1 Structure	15
2.2.2 Function	15
2.2.3 Fibrillogenesis *in vitro*	15
2.2.4 Mechanical Properties	17
2.3 Laminin (LM)	17
2.3.1 Structure	17
2.3.2 Function	18
2.3.3 Polymerization *in vitro*	19
II Material and Methods	**20**
3 Micropillar Substrates	**21**
3.1 Silicon Micropillar Arrays	21
3.1.1 Fluorosilane Coating	22
3.2 Poly(dimethylsiloxane) (PDMS) Microarrays on Glass Coverslips	25
3.3 Polyurethane (PU) Microarrays on Glass Coverslips	25
3.4 PDMS Microarrays on Elastomeric Silicone	25
4 Production of ECM Nanofibrils	**27**

4.1	Isolation of Plasma Fibronectin	27
4.2	Western Blotting of Purified Fn	28
4.3	Surface Activity of ECM Proteins	28
4.4	Fluorescent labeling of Fn	29
4.5	ECM Nanofibrils on Silicon Micropillar Arrays	29
4.6	SEM Analysis of Fn Nanofibril Diameter	29
4.7	Fn Nanofibrils on PDMS Micropillar Arrays	30
4.8	Fibronectin (Fn) Nanofibrils on PU Micropillar Arrays	30
4.9	Fn Nanofibrils on Stretchable PDMS Micropillar Arrays	31

5 Cell adhesion to Extra Cellular Matrix Nanofibrils   33

5.1	Transfer of Fn onto Polyethyleneglycol (PEG) Hydrogels	33
	5.1.1 Production of PEG Hydrogels	33
	5.1.2 Covalent Grafting of Nanofibrils onto PEG Hydrogels	33
5.2	Cell Adhesion to Fn and LM on Polyethyleneglycol Hydrogels	34
5.3	Immunostaining of ECM Proteins	35
5.4	Staining of Focal Adhesions on LM and Fn nanofibrils	35

6 Förster Resonance Energy Transfer (FRET) Analysis of Fn   36

6.1	FRET labeling of Fn	36
	6.1.1 Thiol-reactive Acceptor Labeling	36
	6.1.2 Amine-reactive Donor Labeling	37
6.2	Western Blotting of Labeled Fn	37
6.3	Determination of Degree of Labeling	37
6.4	FRET Calibration to Molecular Unfolding of Fn	38
	6.4.1 Circular Dichroism (CD) Spectroscopy	38
	6.4.2 FRET Signal During Unfolding of Fn	39
6.5	FRET Measurements of Fn Surface Films	39
6.6	FRET Measurements of Fn Nanofibrils	41
	6.6.1 FRET Imaging of Fn Nanofibrils	41
	6.6.2 Data analysis	41

7 Mechanical Properties of Fn Nanofibrils   42

7.1	Extensibility of Fn Nanofibrils	42
7.2	Atomic Force Microscopy (AFM) of Fn nanofibrils	42
	7.2.1 Sample Preparation	42
7.3	Scanning Electron Microscopy (SEM) of Probed Fn nanofibrils	44
7.4	Data Analysis of AFM Experiments	44

III   Results and Discussion   45

8   Surface Activity of ECM Proteins   47

8.1	Protein Accumulation at the Air-Buffer Interface		47
8.2	Discussion		47
	8.2.1	Dynamics of Protein Accumulation at the Air-Buffer Interface	47
	8.2.2	Structural Consequences of Protein Accumulation	50

9 Size Control of ECM Nanofibrils   52

9.1	Silicon Microarrays Produced by Reactive Ion Etching (RIE)		52
9.2	Fn Nanofibrils on Micropillar Substrates		53
	9.2.1	Discussion	55
9.3	Fabrication of LM and COL Nanofibrils		58
	9.3.1	LM Nanofibrils	58
	9.3.2	COL I Nanofibrils	59
	9.3.3	Western Blot Analysis of ECM Preparations	60
	9.3.4	Nanofibrils Consisting of Other Biopolymers	61
	9.3.5	Discussion	61
9.4	Cell Adhesion to Fn and LM Nanofibrils		62
	9.4.1	Discussion	63

10 FRET of Fn Nanofibrils   65

10.1	Calibration of Fn FRET Probe		66
	10.1.1	Circular Dichroism (CD) Spectrometry of Fn Unfolding	66
	10.1.2	Wavelength Determination for FRET Measurements	66
	10.1.3	FRET as a Function of Fn Unfolding	66
	10.1.4	Discussion	70
10.2	Determination of Ratio between FRET Labeled and Unlabeled Fn		71
10.3	FRET of Fn Surface Films		72
	10.3.1	Discussion	73
10.4	Fn Nanofibrils on PU		74
	10.4.1	Discussion	74
10.5	Fn Nanofibrils on Stretchable Substrates		76
	10.5.1	Discussion	76

11 Mechanical Properties of Fn Nanofibrils   80

11.1	Extensibility of Fn Nanofibrils		80
	11.1.1	Discussion	82
11.2	AFM and SEM Analysis of Fn Nanofibrils		83
	11.2.1	Experimental Approach	83
	11.2.2	AFM Analysis	85
	11.2.3	SEM Analysis and Resulting Bending Moduli	85
11.3	Discussion		88
	11.3.1	Comparison of Results to Related Biopolymers	88
	11.3.2	Sources of Error	89

IV Conclusions and Outlook — 90

12 Protein Structure at Interfaces and Within Nanofibrils — 91
    12.1 Conclusions — 91
    12.2 Outlook — 91
        12.2.1 Use of Protein Films — 92

13 AFM Investigation of Fn Fibrillogenesis — 93
    13.1 Conclusions — 93
    13.2 Outlook — 93

14 Production of Polymer Nanofibrils on Micropillar Arrays — 96
    14.1 Conclusions — 96
    14.2 Outlook — 96

List of Figures — 99

Bibliography — 101

A Appendix — 111
    A.1 Abbreviations — 111
    A.2 Immunofluorescence Stain of COL and Fn Nanofibrils — 113
    A.3 Laser Stability Test — 114
    A.4 FRET Calibration in Solution — 115
    A.5 Photobleaching during FRET image acquisition — 116

Abstract

All cells in our body are surrounded by Extra Cellular Matrix (ECM), from which they derive biochemical, structural and mechanical signals. One of the main fibrillar ECM protein components is Fibronectin (Fn), which is believed to act as a mechanochemical signal transducer. A current hypothesis is that Fn can undergo structural transitions upon stretching, which can alter Fn binding site accessibility and ultimately lead to an adapted cell response.

While this hypothesis has existed for several years, the lack of suitable model systems prevented its proof. The aim of this work was to (i) produce regular arrays of Fn nanofibrils, (ii) control the alignment, diameter and tensile state of those nanofibrils, and (iii) to determine their structural and mechanical properties.

During this work, a new method to create regular arrays of Fn nanofibrils was developed. This method allows the control of nanofibril directionality and diameter and can also be used to produce nanofibrils from other ECM proteins, such as Laminin (LM) and Collagen (COL). The method depends both on a protein's ability to accumulate at the air-buffer interface and its ability to self-associate. The production of nanofibrils from various polymers that share these properties is thus possible.

The resulting nanofibrillar arrays can be produced on a variety of mirostructured materials, ranging from Silicon over Poly(dimethylsiloxane) (PDMS) to Polyurethane (PU). The biofunctionality of different ECM nanofibrillar arrays was demonstrated by specific cell adhesion after nanofibril transfer onto non-fouling Polyethyleneglycol (PEG) hydrogels.

An investigation of both the molecular structure and the mechanical properties of Fn nanofibrils was performed by Förster Resonance Energy Transfer (FRET) and Atomic Force Microscopy (AFM) experiments.

Fn molecules form a surface film after application of Fn into a drop of Phosphate Buffered Saline (PBS). FRET analysis of Fn was performed to determine the degree of Fn molecular unfolding. It could be shown that Fn within surface films only unfolds upon surface de-wetting, which coincides with nanofibril formation. The produced nanofibrils show an elongation at break of 200 %. Ruptured nanofibrils retract to 30 % of their original length, but the Fn molecules within nanofibrils do not re-fold completely, as derived from FRET measurements. The pre-strained Fn nanofibrils display a high effective Young's modulus of E $\approx$ 0.1 - 6 GPa, as determined by AFM experiments.

In summary, the production, control and characterization of novel ECM models was accomplished in this work, which can be used to investigate cell adhesive response.

Part I
Introduction

Chapter

1

Mechanosensing in the Extra Cellular Matrix

In every organ, the space between cells is filled with macromolecular assemblies of proteins and polysaccharides, which are summarized under the term Extra Cellular Matrix (ECM). Cells locally secrete and assemble ECM components into a meshwork, which they continuously remodel afterwards [1]. This interaction of cells with their ECM is central to processes occurring in a multicellular organism, ranging from early development over wound healing to malignant processes such as tumor outgrowth [2].

Early experimental work focused on the molecular composition of both the ECM and the cellular receptors binding to it. During the second half of the 20th century, major ECM protein components, such as Fibronectin (Fn), Laminins (LMs) and Collagens (COLs) were identified as well as their cellular binding partners, the integrin family of receptors [3].

However, recent studies have shown that the cell-matrix interaction is regulated not only by the specific binding site repertoire of the matrix, but also by its structural and mechanical properties. Recently, Christopherson and co-workers showed that laminin-coated nanofibrils with average diameters of 283 nm favored differentiation of rat hippocampal-derived Neural Stem Cells (NSCs) into glial cells, whereas the same cells developed into the neuronal lineage on nanofibrils with a mean diameter of 749 nm [4].

In a seminal piece of work, Engler and co-workers [5] could show that Mesenchymal Stem Cells (MSCs) differentiate into distinct cell types based on the elasticity of a collagen coated cell substrate.

Parallel to these observations, potential mechano-chemical signal converters have been identified [6], which are either cellular proteins or, as in the case of Fn, ECM components. A schematic view of the cell-ECM link is given in Figure 1.2.

On a molecular level the conversion of mechanical signals into a biochemical and cellular response involves the unfolding of tertiary and secondary protein structure [6], thereby altering the binding site accessibility of the mechanosensor. While force-induced unfolding of Fn has been convincingly demonstrated by Vogel and

Figure 1.1: Extra Cellular Matrix (ECM) Rigidity of Different Tissues. When seeded on matrices of distinct elastic modulus, Mesenchymal Stem Cells (MSCs) will differentiate into cell types corresponding to the tissue elasticity found *in vivo*. Reproduced from [5].

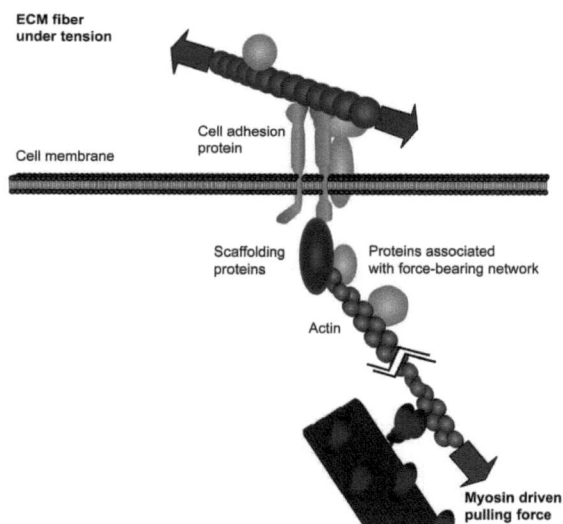

Figure 1.2: Model of Cell-ECM Interaction. The molecular motor myosin (red) pulls on an actin fiber thereby applying force to a protein network (green) that physically links the cytoskeleton to the extracellular matrix. Diverse proteins that are associated with the various molecules of the force-bearing network are given in yellow. Reproduced from [6].

co-workers [7], its role as a mechanosensor could not be elucidated experimentally. Molecular dynamics simulations suggest that Fn unfolding leads to a switch in binding site accessibility towards cell receptors [8], but this has not been demonstrated experimentally.

The aim of the present work was thus to develop an artificial substrate consisting of a regular array of Fn nanofibrils, which could be stretched and relaxed in order to alter the degree of unfolding of Fn molecules forming the nanofibrils. Based on experiments in the group of Prof. Viola Vogel (ETH Zürich) [9, 7, 10, 11, 12], the Förster Resonance Energy Transfer (FRET) signal of dual-labeled Fn was chosen as an indicator of molecular unfolding within Fn nanofibrils. Since cell response is not only dependent on the structural but also on the mechanical properties of the cell environment, Atomic Force Microscopy (AFM) experiments were conducted to derive the effective Young's Modulus of the produced nanofibrils.

Chapter

2

ECM Proteins

In this chapter, I will summarize key properties of the ECM proteins used in this work and review existing methods to produce ECM fibrils *in vitro*.

2.1 Fibronectin (Fn)

Fn was first isolated as "cold-insoluble globulin" from blood plasma cryoprecipitate [13] and later shown to be one of many splice isoforms of a "large cell surface protein" [14], which is lost upon oncogenic transformation of fibroblast cells [15]. Its vital role in animals is underlined by the observation that mouse embryos lacking Fn die at an early stage [16].

While the functional properties of both cellular Fn (cFn) and plasma Fn (pFn) appear to be largely similar, it has to be stated that cFn can additionally contain EIIIA and EIIIB modules, which are assumed to convey distinct cell binding abilities to cFn [2].

Secreted as a dimer by liver hepatocytes, pFn is found at concentrations of ∼300 mg ml^{-1} [2] and the purified protein is commercially available from human or bovine source. This product is commonly used as a cell-adhesive substrate coating for research use. Since the Fn preparation used in this work was derived from blood plasma, I will restrict the following discussion to pFn.

2.1.1 Structure

The pFn monomer is encoded by a single gene, resulting in a glycoprotein with an apparent molecular weight of 230-250 kDa. Plasma Fn is a dimeric molecule that consists of more than 50 repeats of type FnI, FnII, and FnIII. All the repeats are composed of β-sheet motifs, and representative crystal structures are given in the upper panel of Figure 2.1.

The solution structure of pFn is a backfolded disk, which is stabilized by ionic interactions [17] (see Figure 2.2). Cells bind to this soluble Fn via integrin receptors and assemble it by a force-dependent mechanism into nanofibrils with diameters

Fibronectin (Fn)

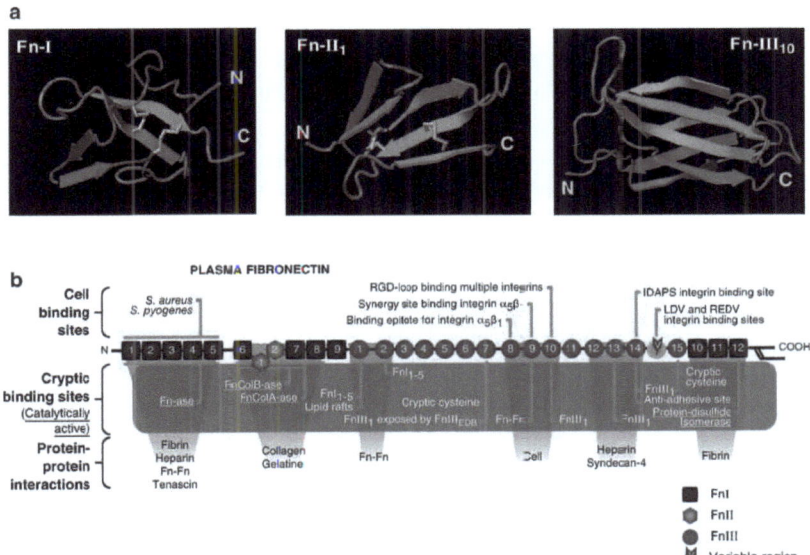

Figure 2.1: Fn structure. (a) Three types of modules exist within Fn. Both FnI and FnII are stabilized by two internal disulfide bonds, shown in yellow. In contrast, FnIII modules which form the central part of the Fn monomer, including the cell binding site, are not stabilized by covalent bonds. (b) Fn contains a large number of molecular recognition and cryptic sites, including the cell binding site RGD; the synergy site PHSRN, which is recognized by α5β1 integrins; and the sequence IDAPS in the HepII region of Fn that supports α4β1-dependent cell adhesion. The cryptic sites include various Fn self-assembly sites whose exposure is needed to induce Fn fibrillogenesis. Modified from [8].

between 10 and 1000 nm [18]. While the requirement of cell contractility has been shown and unfolding of Fn absorbed to the surface of cells has been reported, it is still undetermined:

- which molecular processes occur during Fn fibrillogenesis,
- how Fn modules interact within Fn nanofibrils, and
- of what type the stabilizing interactions within Fn fibrils are.

Recently, a model for the Fn fibril structure was proposed, which is based on data derived from nuclear magnetic resonance (NMR) experiments [17] (see Figure 2.2).

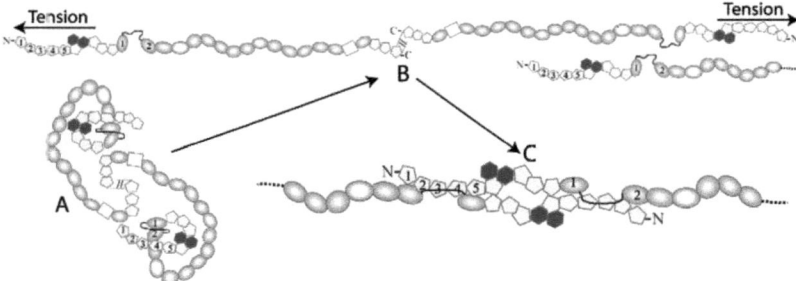

Figure 2.2: Proposed Model of Fn Fibrillar Interactions. (A) Fn exists in a globular, soluble state in plasma. The interdomain interactions defining this state are disrupted by cell-generated tension (B). (C) The N-termini of two extended Fn molecules form a tight complex through the FnI_{1-5} : $FnIII_{1-2}$ interaction, thereby creating Fn protofibrils. Reproduced from [17].

2.1.2 Function

Cell Adhesion

Fn allows cell attachment and spreading via its several integrin binding sites. Cell-ECM contact points, where co-alignment of Fn fibrils and actin filaments is observed [19], were initially termed "fibronexi", a term nowadays specifically used for large cell-ECM contacts found on tissue myofibroblasts [20]. Fn is among the most frequently used cell-adhesive coatings that support cell attachment [21].

Cell Morphology

Upon addition of pFn to oncogenically transformed cells, which lack Fn at the cell surface, restoration of microfilament architecture and cell morphology was reported [22]. This demonstrates a function of Fn in outside-in signaling. Conversely, the presence of cytochalasin B, which disrupts actin microfilaments, leads to the release of surface Fn into the medium [23], indicating that cell-ECM contacts are force-sensitive structures.

Cell Migration

The migratory response to Fn depends on the cell type. Most cells, including fibroblasts, smooth muscle cells, bone cells and neural crest cells, migrate along gradients of Fn, while the migration of liver epithelial cells is inhibited [2]. One of the best studied examples of pFn function is its role in wound healing. The sequence of events seems to be that, soon after wounding, Fn and fibrin appear in the area of the wound. These proteins then serve as a substrate for adhesion and migration of the cells repairing the defect and, in most cases, subsequently disappear [2].

Differentiation and Proliferation

While it is difficult to distinguish between effects on adhesion and effects on proliferation and cell differentiation *per se*, numerous studies indicate that Fn often promotes proliferation and cell-type specific differentiation. Since Fn forms part of a complex ECM, which also harbors growth factors and other mitogens, it has so far been impossible to conclude that these are direct effects exerted by Fn or any of its fragments.

2.1.3 Fibrillogenesis *in vitro*

Reported methods to produce Fn fibrils are:

1. the manual pulling of single fibrils out of a concentrated drop of Fn [11, 24, 25], yielding fibrils with diameters between 0.5 µm and 5 µm [11] (see Figure 2.3)

2. the self-assembly of Fn networks underneath a lipid monolayer followed by gradual expansion of the film [26], yielding fibrils with several microns in diameter

3. surface-induced fibrillogenesis on hydrophobic substrates [27], yielding nanofibrils with diameters between 20 nm and 100 nm

4. the spontaneous aggregation of Fn into fibrils via re-folding of Fn during dialysis [28], yielding nanofibrils between 5 nm and 20 nm, or

5. the addition of anastellin, an aggregation-inducing fragment of Fn [29], which leads to thickening of Fn fibrils produced by cells [30].

The drawbacks of these approaches are that method 1. can only yield few individual fibrils and the precise control of fiber alignment and diameter is not possible for the remaining methods.

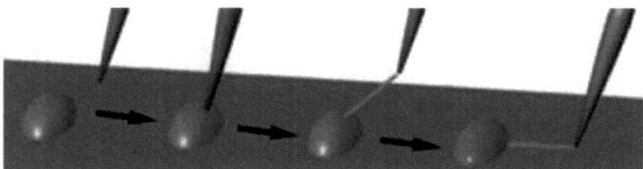

Figure 2.3: Fibrils Pulled Out From Fn Solution. Fn fibrils pulled out of a drop of Fn solution. A pipette tip is submersed slowly into a concentrated solution (0.76 mg/mL) of Fn and removed to generate polymerized Fn fibers. Modified from [11].

2.1.4 Mechanical Properties

Molecular unfolding

The first experimental AFM study investigated pFn physisorbed to a petri dish [31]. In this study, unfolding of all three types of modules was distinguishable by differences in unfolding step size and resulted in a dimeric molecule being extended from a countour length of less than 140 nm to over 1700 nm. By fitting the force-distance curve of Fn unfolding events to a worm-like chain model, the persistence length of Fn was estimated to be 0.3 ± 0.2 nm. Further AFM experiments concentrated on the unfolding of individual FnIII modules, which suggested a hierarchy of unfolding strengths for the various modules [32], as supported by Steered Molecular Dynamics (SMD) simulations [33]. The results of these studies are summarized in Figure 2.4.

An additional level of complexity is introduced by the observation that unfolding of the RGD-containing and cell-binding $FnIII_{10}$ module occurs via different intermediate states, which are dependent on the pulling velocity [34, 35, 36].

These findings are based on pulling forces acting from the N-terminus to the C-terminus of the molecule. However, a recent SMD investigation, where the pulling force was acting on the RGD-site within the cell binding module, found a single intermediate state [37].

Fibril Extensibility

The extensibility of Fn fibrils has been investigated both in cell environments [38, 9, 10, 39] and in completely artificial systems [11]. These studies showed that fibrils

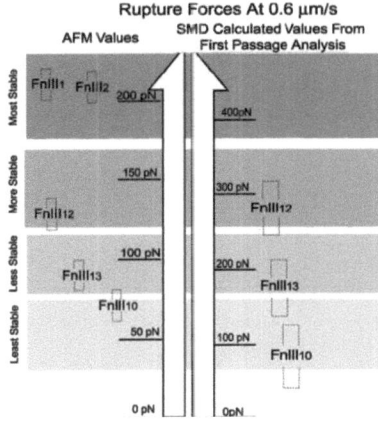

Figure 2.4: Fn Domain Unfolding Hierarchy. Boxes shown surrounding each module indicate error bars from AFM or SMD results (3.0 ± 0.5 Å). Progressively grayer bars indicate increasing mechanical stability. AFM results are taken from Oberhauser, et al. [32]. Modified from [33].

can shorten down to 25 % of their original length when detached from the cellular surface [38].

In the past decade, extensive work in the lab of Prof. Viola Vogel used Förster Resonance Energy Transfer (FRET) to correlate the extension of Fn fibrils with the structural properties of Fn molecules that form them.

Using the FRET labeling scheme described in Figure 2.5, it could be shown that:

- Fn in ECM fibrils is partially unfolded [10, 9, 7],
- cells progressively unfold Fn in these ECM fibrils [39],
- Fn fibrils do not re-compact into the solution form [10, 11], and
- the extensibility of artificial Fn fibrils lies between 600 % and 1200 % [11]

2.2 Collagen (COL)

There are more than 30 Collagens (COLs) and COL-related proteins but the most abundant are Collagens I and II that exist as D-periodic (where $D = 67$ nm) fibrils. The fibrils are of broad biomedical importance and have central roles in embryogenesis, arthritis, tissue repair, fibrosis, tumor invasion, and cardiovascular disease [44].

Collagens I and II spontaneously form fibrils *in vitro*, which shows that COL fibrillogenesis is a self-assembly process. However, the situation *in vivo* is not that simple; COL I-containing fibrils do not form in the absence of Fn, Fn-binding and COL-binding integrins, and COL V [44].

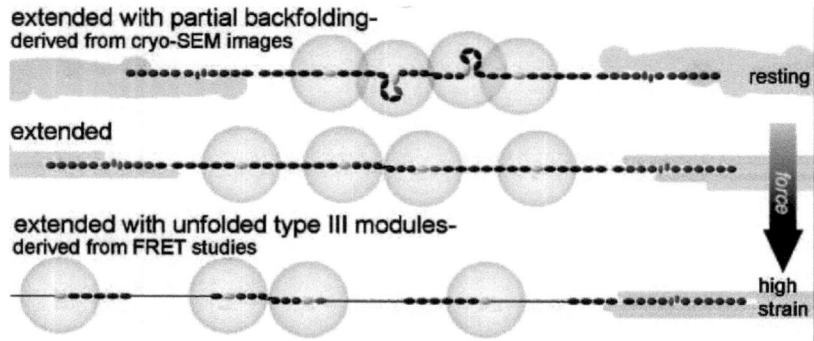

Figure 2.5: Putative Fn Conformations within ECM Fibrils. Fn consists of tandem repeats of type I (dark blue ovals), II (narrow, light blue ellipses), and III modules (dark red ovals). Average end-to-end lengths of each module type are drawn to scale using lengths of 2.5 nm for Fn type I [40], 0.7 nm for Fn type II [41], and 3.2 nm for FnIII modules [42]. Two free cysteines are present on each monomer within FnIII$_7$ and III$_{15}$ (yellow). Energy transfer between donors and acceptors bound to free cysteines are limited to within approximately double the Förster radius (12 nm), denoted by gold circles around III$_7$ and III$_{15}$. High-resolution cryoscanning electron microscopic images of Fn fibrils [28, 43], taken together with FRET studies, suggest that fully relaxed fibers do not contain the compact quaternary structure, but are composed of Fn in an extended conformation with partial backfolding of its arms upon themselves (nodules; top). Cell-generated tensile forces first extend Fn fibrils (middle) and finally unfold FnIII modules (bottom). Modified from [10].

COL I can be easily purified in a solution of acetic acid from calf skin or rat tail and is commercially available for the generation of 2D and 3D cell substrates. Being most intensely studied and subject of the experiments described later, I will restrict the following discussion to COL I.

2.2.1 Structure

COL I comprises three polypeptide chains ($\alpha 1(I)_2 \alpha 2(I)$) and is only assembled into fibrils after extracellular cleavage of the procollagen triple-helix, also called Tropocollagen (TC) [45].

For the three chains to wind into a triple helix, they must have the smallest amino acid, glycine, as every third residue along each chain. Each of the three chains therefore has the repeating structure Gly-Xaa-Yaa, in which Xaa and Yaa can be any amino acid but are frequently the iminoacids proline and hydroxyproline. An overview of COL I fibrillogenesis is given in Figure 2.6.

2.2.2 Function

COL I makes up 90 % of all COLs and is found throughout the body except in cartilaginous tissues [45]. It serves an important role as a structural component of the ECM, specifically in tendons, skin and the cornea. At this point, biomaterials based on COL I are in clinical use to accelerate wound healing and bone regeneration.

The main integrin binding site, GFPGER, plays a role in angiogenesis, endothelial cell activation, and osteoblast differentiation [47]. Recently, a map of interaction sites of COL I has been published, where, among others, a matrix metalloprotease sensitive site and a Fn binding site are included [47]. This indicates that ECM remodeling is a complex process, which involves several ECM components at any instance.

2.2.3 Fibrillogenesis *in vitro*

Extensive studies in the past decades could show that fibrillogenesis of COL I is a spontaneous, entropy-driven self-assembly process [44, 45].

After isolation from tissue, COL I is usually suspended in acidic buffer (pH $\leq$ 2). Upon reconstitution in buffer of pH increasing up to 8.9 and warming to temperatures between 20°C and 34°C, the Tropocollagen (TC) molecules self-assemble into D-periodic fibrillar structures over the course of several hours. At 34°C, fibril diameters are typically in the range between 20 nm and 70 nm. Lower temperatures generally result in broader fibrils, with diameters of up to 200 nm found at 20°C [45].

The ionic composition of the buffer also plays an important role, since the characteristic D band pattern only forms in the presence of Potassium ions [48]. In recent

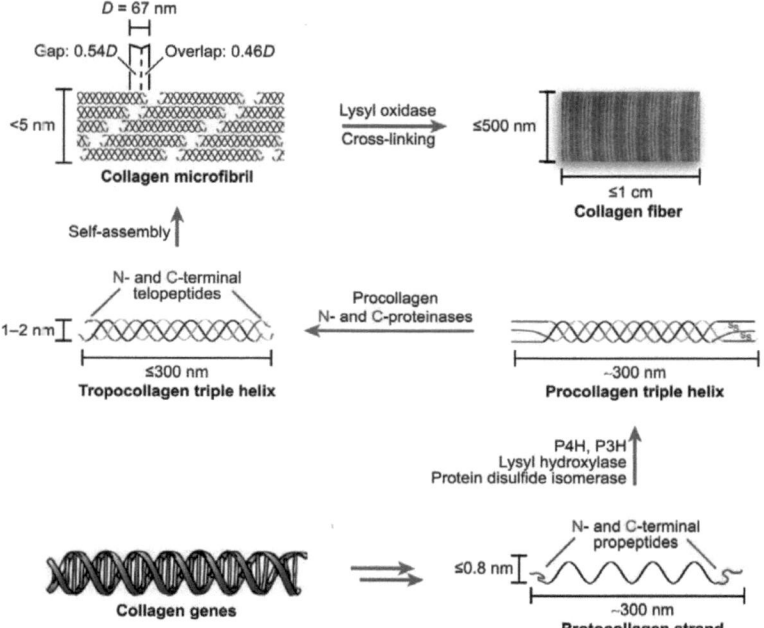

Figure 2.6: COL Fibrillogenesis. Procollagen consists of a 300-nm-long triple-helical domain (comprised of three α-chains each of approx. 1000 residues) flanked by a trimeric globular C-propeptide domain and a trimeric N-propeptide domain. Procollagen is secreted from cells and is converted into Tropocollagen (TC) by the removal of the N- and C-propeptides by procollagen N-proteinase and procollagen C-proteinase respectively. The TC triple-helix generated in the reaction spontaneously self-assembles into cross-striated fibrils. The fibrils are stabilized by covalent cross-linking, which is initiated by oxidative deamination of specific lysine and hydroxylysine residues in COL by lysyl oxidase. Reproduced from [46].

years, this knowledge could be used to fabricate highly ordered COL matrices [48] and investigate the structural [49, 50] and mechanical [51] properties of COL fibrils.

Another method to produce COL nanofibrils is electrospinning, where a positively charged capillary filled with COL solution ejects a jet of polymer solution onto a grounded target [52].

2.2.4 Mechanical Properties

The elastic modulus of a TC monomer is estimated to lie in the range of E = 6-7 GPa [53, 54], whereas AFM measurements show that dehydrated fibrils of COL I from bovine Achilles tendon [55] and rat tail tendon [56] have E ≈ 5 GPa and E ≈ 11 GPa, respectively.

Because COL fibrils are anisotropic, the shear modulus is also an important measure of the strength of a Collagen fibril. AFM experiments recently revealed that dehydrated fibrils of COL I from bovine Achilles tendon have a shear modulus of G = 33 MPa [57]. Hydration of these fibrils reduced their shear modulus significantly, whereas carbodiimide-mediated cross-linking increased their shear modulus.

It is noteworthy that a certain level of cross-linking is favorable for the mechanical properties of COL fibrils, but excessive cross-linking results in extremely brittle COL fibrils [53], a common symptom of aging. An analysis by Buehler [53] of the mechanical properties of COL fibrils suggests that nature has selected a length for the TC monomer that maximizes the robustness of the assembled COL fibril via efficient energy dissipation. Simulations indicate that TC monomers either longer or shorter than 300 nm (which is the length of a COL I triple helix) would form fibrils with less favorable mechanical properties.

2.3 Laminin (LM)

The members of the LM family are major constituents of all Basement Membranes (BMs), sheet-like extracellular structures, present in almost all organs. The LMs bind to cell surface receptors and thereby tightly connect the basement membrane to the adjacent cell layer. This provides for the specific basement membrane functions to stabilize cellular structures, to serve as effective physical barriers, and furthermore, to govern cell fate by inducing intracellular signaling cascades. Many different types of diseases involve basement membranes and LMs. Metastasizing solid tumors must pass through basement membranes to reach the vascular system, and various microbes and viruses enter the cells through direct interaction with LMs [58].

2.3.1 Structure

LMs are composed of three different polypeptide chains, termed α, β and γ. At present, 5 α, 3 β, and 3 γ chains are known for mouse and human [59]. All chains are glycosylated, and a few chains have been shown to have glycosaminoglycan side chains. To differentiate between the LM heterotrimers, the chain composition is stated in arabic letters, e.g. LM-531 consists of the α-5, β-3, and γ-1 chain.

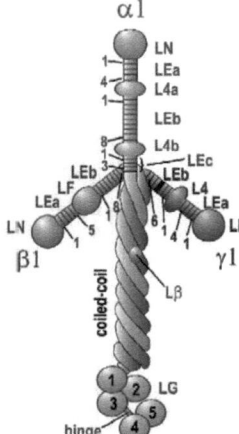

Figure 2.7: LM-111 Structure. Abbreviations used: LN, laminin N-terminal domain; LE, laminin epidermal growth factorlike repeats; L4, laminin 4 domain; LF, laminin four domain; and LG, laminin globular domain. Modified from [59].

LM-111, can be isolated in high yield and purity from Engelbreth-Holm-Swarm (EHS) tumor [60]. It is widely used as an adhesive coating for neuronal cell substrates and was the LM isoform used in the present work. The following discussion will thus concentrate on findings related to EHS LM.

2.3.2 Function

A major feature of most LMs is their ability to form networks via the globular LN domains (laminin N-terminal domain, see Figure 2.7). Together with other BM components, LMs form the largest polymers in the body, as they form continuous sheets, for instance skin, with a single basement membrane polymer reaching from head to toe under the epidermis [58].

Cell binding to LM occurs via integrin receptors to the LG domains (laminin globular domains, see Figure 2.7). Apart from cell adhesion and linkage of the cytoskeleton to the ECM, LM-binding integrins act as signalling receptors, mediating growth, differentiation, and survival signals from the ECM [61].

Tsiper et al. [62] revealed that Schwann cells, which myelinate axons in the Peripheral Nerve System (PNS), form fibrils when supplied with EHS LM. While a role for LM fibrils in Schwann cell guidance during development was suggested by the authors, neither the mechanism of LM fibrillogenesis, nor the physiological relevance of these structures has been elucidated.

2.3.3 Polymerization *in vitro*

Polymerization is an integral part of the biological function of LMs and initiates BM assembly. Other components then assemble close to the LMs. LM polymerization is calcium dependent, where the LN domain (laminin N-terminal domain, see Figure 2.7) of each chain noncovalently binds other LN domains so that three chains meet. Yurchenco *et al.* could follow the LM polymerization process using transmission electron microscopy, where they found that a chilled solution of LM-111 would gradually polymerize when increasing the temperature to 34 °C (Figure 2.8).

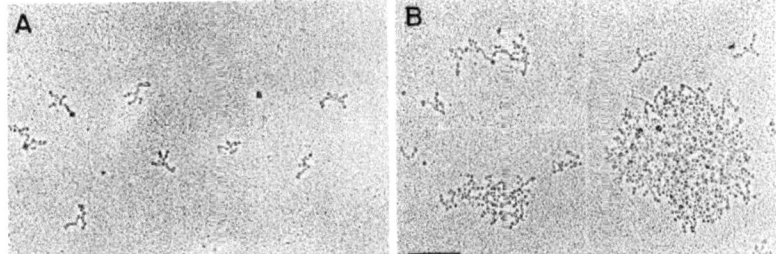

Figure 2.8: LM Polymerization. Rotary shadow platinum replicas of LM maintained at different temperatures LM, in neutral phosphate buffer at 0.6 mg ml^{-1} was either maintained on ice (A) or incubated at 35 °C (B) for 1 h. The sample was then diluted to 5 pg ml^{-1} in ammonium acetate/glycerol buffer, sprayed onto mica, and prepared for rotary shadowing. Panel B is a composite to show typical areas of LM complexes. Scale bar: 200 nm. Reproduced from [63].

Part II

Material and Methods

Chapter

3

Micropillar Substrates

Four different types of micropillar substrates were used during this work:

- Silicon micropillars
- Poly(dimethylsiloxane) (PDMS) micropillars on glass coverslips
- Polyurethane (PU) micropillars on glass coverslips
- PDMS micropillars on elastic silicone sheets

All structures were produced using standard photolithography and soft moulding techniques. The microarray geometry is defined by the parameters depicted in Figure 3.1.

3.1 Silicon Micropillar Arrays

Silicon wafers (2 inch, orientation 100, p-type, #S4181) were dried on a hotplate at 200°C for 5 minutes. After cooling down, 1 ml S1818G2 photoresist (Rohm & Haas Deutschland GmbH) was applied and spin coated for 10 s at an acceleration of 100 rpm s^{-2} to a limit of 1000 rpm, immediately followed by an acceleration of 900 rpm s^{-2} to a limit of 5500 rpm for 40 s. After 1 min pre-bake at 115°C, the wafer was illuminated for 2 s through a chromium mask on a SUSS MJB4 Mask Aligner using the G-Band of a 350 W HBO lamp. After 40 s immersion under continuous agitation in S1818G2 developer (Rohm&Haas Germany), the microstructure was shortly immersed in UltraPure water with a resistivity above 18.2 MOhm and blown dry under a nitrogen stream. The microstructure was hardbaked for 15 min at 115°C before transfer into a Plasmalab 80 Plus (Oxford Instruments) reactive ion etcher. The structures were etched at -10°C to a depth ranging between 3 µm and 20 µm at an etch rate of 70 nm per cycle. The etching cycle is summarized in Table 3.1. The remaining photoresist was dissolved in S1818G2 remover (Rohm&Haas Germany) for 5 min, the structures were thoroughly rinsed in UltraPure water. Organic residues were etched away by immersion of the microstructure for 2 h in a 1:3 mixture of

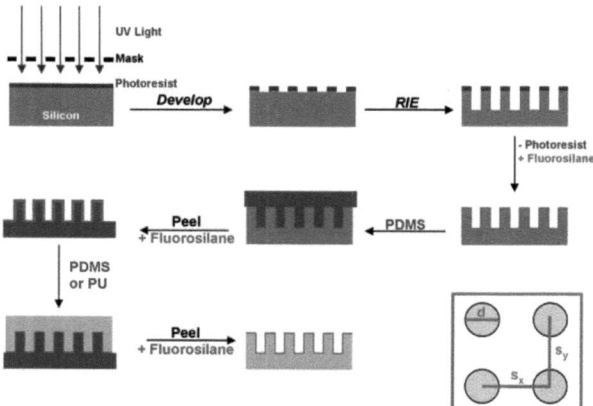

Figure 3.1: Microarray Fabrication Process. A sequence of passivation and moulding steps leads to micropillar structures derived from silicon micropillar arrays. Three parameters define the microarray geometry: pillar diameter (d); center-to-center separation in x-direction (s_x), and center-to-center separation in y-direction (s_y, see inset, bottom right).

Table 3.1: Parameters used for reactive ion etching of silicon microarrays.

Step	t [s]	p [mTorr]	CHF_3 [sccm]	SF_6 [sccm]	P_{RF} [W]	P_{ICP} [W]
Passivation	8	70	50	0	30	100
Etching	5	40	18	15	30	300

30 % v/v H_2O_2 and H_2SO_4. After copious rinsing with UltraPure water, the structure was blown dry under a nitrogen stream and dried at 150°C for 15 min.

3.1.1 Fluorosilane Coating

The microstructure was activated for 1 minute in a Plasma System 100 (PVA TePla) at 1 mbar O_2, 200 W. Vapor deposition of 25 μl tridecafluorosilane (#T2494, UCT) was performed in an desiccator (#FB35005, Fisher Scientific, Germany) at 1 mbar for 30 min.

Silicon Micropillar Arrays

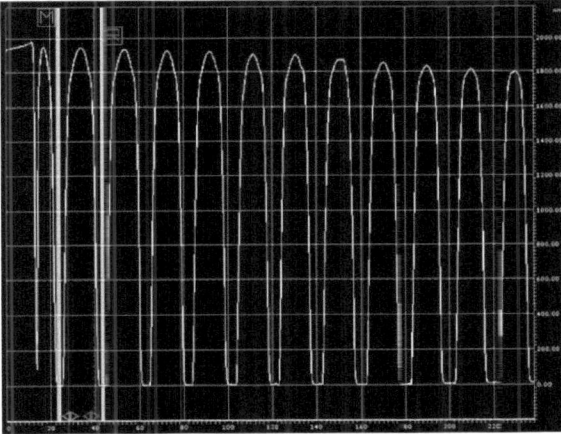

Figure 3.2: Photoresist Thickness Determination. The thickness of the deposited photoresist was measured on a DekTak profilometer. A tip with 5 µm diameter was scanned along the s_x axis of a microstructure with 10 µm pillar diameter and 20 µm center-to-center spacing (see Figure 3.1). The curved pillar tops visible in the scan are caused by convolution of the tip shape with the underlying microstructure. As derived from the length scale on the right hand side, the average film thickness is 1800 nm.

Figure 3.3: Effect of Fluorosilane Coating. The contact angle of a drop of water increases upon silanization (left: untreated, right: silanized). Substrates from top to bottom: Poly(dimethylsiloxane) (PDMS), Silicon, Polyurethane (PU).

3.2 PDMS Microarrays on Glass Coverslips

To produce PDMS pillar arrays, a negative PDMS mask was created using silicon micropillar structures as a casting mould. SYLGARD 184 (#608284, Sasco Holz GmbH) prepolymer and curing agent were mixed at a ratio of 9:1 and degassed in an exsikkator for 60 min. PDMS was cured at 140°C for 60 min and peeled off the wafers. The PDMS negative was rendered non-adhesive analogous to the silanization protocol described above. This negative was then coated with freshly degased PDMS and pressed face-down onto a coverslip (#H878.1, Carl Roth). Curing and peel-off of PDMS was performed as described above.

3.3 PU Microarrays on Glass Coverslips

A photoactivable polymerization solution was prepared based on a recipe by Choi et al. [64] (see Table 3.2). The solution was thoroughly mixed at room temperature in a 50 ml polypropylene tube (#210261, Greiner BioOne, Germany) on a rotisserie until the photoinitiator (Igracure 184) of powder type completely dissolved. Shielded from light, the mixture was stored on a rotisserie at room temperature for up to three months. For micromoulding, the solution was poured onto a PDMS negative

Table 3.2: Composition of PU polymerization solution.

m [g]	Reagent Name	Supplier
14.7	Ebecryl 284	CYTEC
4.4	M3160	MIWON Co., Korea
0.22	Irgacure 184	Ciba, Switzerland
0.22	Darocur 1173	Ciba, Switzerland

and pressed face-down onto a circular glass coverslip. Curing was performed by Ultraviolet (UV) exposure for 30 minutes at 20 mW cm^{-2} in a UVA Cube 100.

3.4 PDMS Microarrays on Elastomeric Silicone

Analogous to the production of PDMS microarrays on glass coverslips, a silanzied PDMS negative was coated with freshly degased PDMS and pressed face-down onto a flexible silicone sheet (0.1"' NRV G/G 40D, SMI). Curing and peel-off was performed as described above. To ensure superhydrophobic properties of stretched silicone sheets, a silanization step was carried out analogous to the one described in Section 3.1.1.

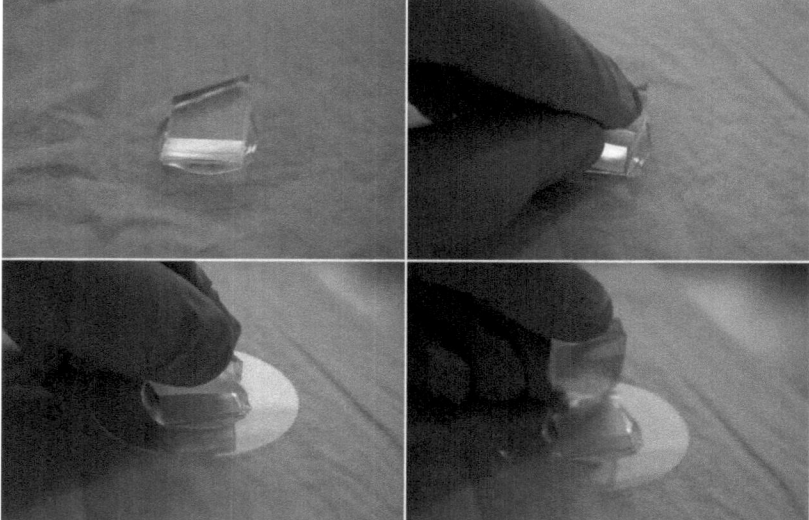

Figure 3.4: Demoulding of PDMS Negative from PU. Starting in the top left image, the process of demoulding is illustrated. The temporal order is clockwise.

Chapter

4

Production of ECM Nanofibrils

In order to produce regular and cell-adhesive ECM nanofibrils, a de-wetting based method was developed, as displayed in Figure 4.1. While this work focused on the production of Fn nanofibrillar arrays, it could be shown that commercially available preparations of COL I and EHS LM are also capable of forming nanofibrils, although at higher protein concentrations.

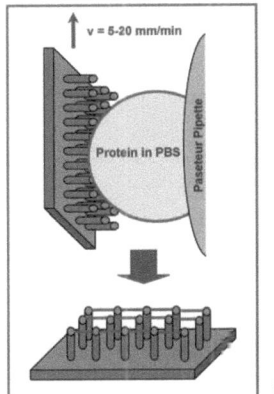 

PDMS pillar microaray on glass coverslip

Figure 4.1: Fabrication of ECM Fibrillar Arrays. The image on the right hand side shows the setup which was used to create the fibrillar arrays. The protein solution was applied on the accessible glass surface of the pasteur pipet (blue arrow).

4.1 Isolation of Plasma Fibronectin

Fn was isolated using a previously described procedure [11]. Frozen human serum (Katharinenhospital Stuttgart, Germany) was thawed at 37°C with agitation. Immediately after thawing, serum was centrifuged in a Sorvall SLA-1500 rotor at 10,600 rpm for 10 min to remove residual red blood cells. To the supernatant, 2 mM Phenylmethanesulphonylfluoride (PMSF) and 10 mM Ethylenediaminetetraacetate (EDTA) were added and the solution was filtered through a 0.4 µm pore filter (Millipore).

Fast Protein Liquid Chromatography (FPLC) columns were prepared according to the manufacturer's protocol. Before applying the protein solution, columns were equilibrated with 5 column volumes of Phosphate Buffered Saline (PBS) containing 2 mM EDTA. The plasma supernatant was passed over a 18 ml Sepharose 4B column (# 4B-200, SIGMA) and the flow through was immediately applied to a 18 ml Gelatin Sepharose 4B column (#17095-601) at a flow rate of 2 ml min^{-1}. The Fn adsorbed to the column was washed with PBS containing 10 mM EDTA and 2 mM PMSF until the OD_{280} of the flowthrough dropped to zero. Fn was eluted from the gelatin column with PBS containing 6 M urea and fractions of 2 ml were collected and the protein concentration determined on a spectrophotometer (Nanodrop 1000, peqlab, Germany).

Fractions were diluted with PBS containing 6 M urea to 2 mg ml^{-1} and frozen at -80°C for long term storage. Aliquots were reconstituted in PBS buffer as needed. For this, PD-10 columns (#17-0851-01, GE Healthcare) were used with PBS as elution buffer according to the manufacturer's protocol.

4.2 Western Blotting of Purified Fn

In order to test if fragmentation of Fn occurred during protein isolation, SDS PAGE and Western Blot analysis were carried out. The concentration of labeled protein fractions was estimated via absorption of the solution at 280 nm and an amount equal to 0.4 µg per lane was loaded onto the gel. Electrophoretic separation and Western Blotting under reducing conditions was performed using 4-20 % polyacrylamide gradient gels according to the manufacturer's protocol (#25244, Pierce). After staining in water containing acetic acid (5 % v/v) and Ponceau S (0.1 % v/v), nitrocellulose membranes were photographed and de-stained in UltraPure water.

4.3 Surface Activity of ECM Proteins

Experiments were performed at room temperature using a NIMA 112D Langmuir-Blodgett trough. After cleaning in 2 % Hellmanex (#634-0442, VWR International), the trough was copiously rinsed with UltraPure water. For all proteins, 140 µg of protein were added drop-wise onto a 70 ml layer of PBS. Surface pressure was recorded over time spans of at least one hour. After 2 hours, the protein film, which had formed on top of the PBS layer, was compressed from 80 cm² to 20 cm² by moving the trough barriers together.

In control experiments, the buffers in which the ECM proteins were dissolved were tested individually on their surface activity. For this, 140 µl of buffer were added drop-wise onto a PBS layer as described above.

4.4 Fluorescent labeling of Fn

Labeling of FN with ATTO 488 maleimide (#28562, Fluka) or ATTO 647 maleimide (#41784, Fluka) was performed according to the manufacturer's protocol.

Shortly, dye was dissolved to 2 mM in Dimethylsulfoxide (DMSO). Fn in 6 M urea was thawed and mixed with an equal volume of PBS containing 8 M Guanidine Hydrochloride (GdnHCl), resulting in a final volume of 2 ml and a Fn concentration of 1 mg ml^{-1}. To this solution, 30 µl of the dye stock solution was added and incubated in the dark for 2 h on a rotisserie at room temperature.

Unreacted dye was separated from protein by size exclusion chromatography using a PD-10 column (#17-0851-01, GE Healthcare) with PBS as elution buffer according to the manufacturer's protocol.

4.5 ECM Nanofibrils on Silicon Micropillar Arrays

In order to improve the spreading of the protein solution, the pipette was activated in an oxygen plasma chamber before use. A piece of duct tape (#4351, tesa Germany) was wrapped around the thick end of a 230 mm pasteur pipette, so that only a 5 mm slit of glass was accessible. The pasteur pipette was then fixed horizontally to a wall. A drop of 100 µl protein solution in PBS was pipetted onto the small slit and the structure to be coated was immediately brought into contact with the droplet. As soon as the meniscus line covered the width of the microarray, the structure was pulled upwards at constant speed. The optimal protein concentration and pulling speed varied depending on the microarray geometry as detailed in Figure 9.4.

EHS LM (#08-125, Millipore, Schwalbach, Germany), COL I from rat tail (COL I (B), #BT-274, Biomedical Technologies, Inc.), COL I from rat tail (COL I (S), #C3867, SIGMA-Aldrich) and COL I from calf skin (COL I (C), #C8919, SIGMA-Aldrich) were employed using the parameters described in Table 4.1.

4.6 SEM Analysis of Fn Nanofibril Diameter

For each parameter set, three independent samples of silicon micropillar arrays were produced on different days. Three areas within each 1 x 2 cm microarray were randomly picked and fibrils were imaged in a Zeiss Ultra 55 Electron Microscope

Table 4.1: Parameters used for the production of Fn nanofibrils.

Protein	d_{pillar} [µm]	c [µg ml^{-1}]	v [mm min^{-1}]
Fn	2.5	12.5	20
Fn	5	25	10
Fn	10	50	5
EHS LM	2.5	25	2
EHS LM	5	50	2
EHS LM	10	100	2
COL I (B)	2.5	50	2
COL I (B)	5	100	2
COL I (B)	10	200	2
COL I (S)	2.5	500	2
COL I (S)	5	500	2
COL I (S)	10	500	2
COL I (C)	2.5	500	2
COL I (C)	5	500	2
COL I (C)	10	500	2

using the InLens detector at an acceleration voltage of 1 kV. Conductive coating of the samples was not necessary. For each sample, 350 nanofibril diameters were extracted using the software provided with the instrument.

4.7 Fn Nanofibrils on PDMS Micropillar Arrays

The parameters used for the production of Fn nanofibrils on PDMS micropillar arrays were identical with those depicted for silicon micropillar arrays given in Table 4.1. After nanofibril production, substrates were wetted with PBS containing 0.1 % w/v MOWIOL-488 (#475904, Calbiochem). Substrate wetting endured 1 h, after which the array was rinsed three times in PBS to remove surfactant.

4.8 Fibronectin (Fn) Nanofibrils on PU Micropillar Arrays

In order to minimize errors in the AFM measurements derived from pillar bending, PU with an elastic modulus of 40 MPa was used instead of PDMS, which generally has an elastic modulus below 10 MPa [65]. The AFM cantilever dimensions additionally required a micropillar geometry which had a center-to-center spacing of 20 µm. Since these PU micropillar arrays tended to wet more readily when contacted by a drop of protein solution, the pulling speed was increased to 20 mm min^{-1}. This led to a more irregular pattern of Fn nanofibrils, which was however sufficient for the analyisis of single Fn nanofibrils.

4.9 Fn Nanofibrils on Stretchable PDMS Micropillar Arrays

The parameters used for the production of Fn nanofibrils on PDMS micropillar arrays were identical with those depicted for silicon micropillar arrays given in Table 4.1.

The substrate was either pre-stretched or relaxed when producing Fn nanofibrils. Array geometry was chosen such that, in both cases, the center-to-center spacing along the fiber pulling direction was $s_x = 20$ µm.

Because of the inherent compression in y-direction when stretching an elastic sheet in x-direction, the pre-stretched substrate displayed a nonuniform center-to-center spacing in y-direction, s_y, which was minimal in the center of the stretched substrate. Effecting a change in the de-wetting behavior of the micropillar array, this led to the occurrence of frequent irregularities within the nanofibrillar array. This effect was reduced when increasing the pulling speed to 20 mm min^{-1}, however at the expense of nanofibril coverage.

After nanofibril production, substrates were wetted with PBS containing 0.1 % w/v MOWIOL-488 (#475904, Calbiochem). Substrate wetting endured 1 h, after which the array was rinsed three times in PBS to remove surfactant. For the analysis of the elongation at break of Fn nanofibrils as well as for FRET experiments on stretched nanofibrils, the substrate was subsequently mounted on a stretching device, depicted in Figure 4.2

Figure 4.2: Stretching Device Used for Fn Nanofibils. The device consisted of two clamps attached to two movable stages, which were controlled by the interface and software provided by the manufacturer (Physik Instrumente). This allowed substrate extensions of up to 400%.

Chapter

5

Cell adhesion to Extra Cellular Matrix Nanofibrils

5.1 Transfer of Fn onto Polyethyleneglycol (PEG) Hydrogels

5.1.1 Production of PEG Hydrogels

PEG diacrylate with a mean molecular weight of 10 kDa (1 g) was mixed with UltraPure water (1655 μl) 2-carboxyethylacrylate (132.5 μl, #552348, Sigma-Aldrich) and 2-hydroxy-4'-(2-hydroxyethoxy)-2-methylpropiophenone (150 μl, #410896, Sigma-Aldrich). After de-gassing and purging with nitrogen, the mixture was kept under nitrogen atmosphere until use. Hydrogels were cast in custom-made chambers which consisted of a bottom silicon wafer that had 0.17 mm glass spacers on it. An objective slide was put on top of the spacers and the prepared PEG 10 kDa solution was applied. To facilitate flow of the PEG solution into the casting chamber, the wafer and slide were activated for 1 minute in a Plasma System 100 (PVA TePla) at 1 mbar O_2, 200 W. UV curing was performed in an UVA Cube 100 at 20 mW cm^2. Swelling of the hydrogels in UltraPure water resulted in detachment from the casting chamber.

5.1.2 Covalent Grafting of Nanofibrils onto PEG Hydrogels

For the covalent attachment of ECM nanofibrils, hydrogels were activated for 60 minutes face-down in a drop of solution containing EDC (78 mg, #153-0990, Bio-Rad) and NHS (19.5 mg, #56480, Fluka) in MES buffer (1000 μl, 1M, pH 5.5). Silicon microstructures containing nanofibril arrays were wetted in PBS containing MOWIOL 4-88 (0.1 % w/v, #475904, Calbiochem), rinsed with PBS and the array was put onto the activated surface of the hydrogel. This setup was enclosed between two objective slides and fixed with paper clamps. After 15 min, the hydrogel was brought into contact with the second array. The reaction was quenched by incubating the hydrogel in PBS containing Bovine Serum Albumin (BSA) (1 % w/v, #11931, SERVA) for 1 h. After rinsing in PBS, the hydrogel was glued into a 35 mm petri dish using Nexaband (#5297, Abbott, Chicago, Illinois) tissue glue. Prior to use, the hydrogel was stored in PBS and equilibrated for at least 2 hours in cell culture medium.

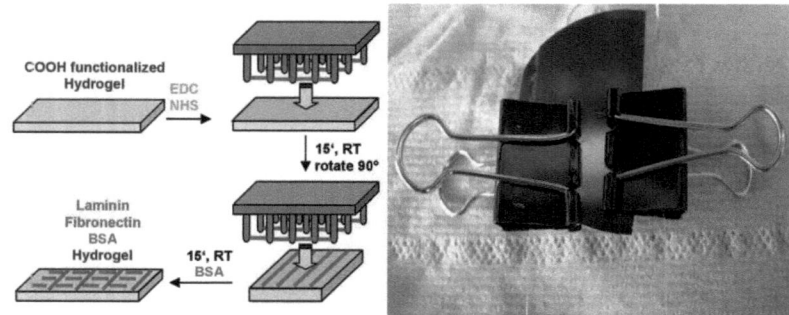

Figure 5.1: Transfer and covalent attachment of LM (red) and Fn (green) nanofibrils onto PEG hydrogels. Hydrogels were copolymerized with 2-carboxyethylacrylate to introduce -COOH functional groups. Standard 1-Ethyl-3-[3-dimethylaminopropyl]carbodiimide hydrochloride (EDC)/N-hydroxysuccinimide (NHS) coupling was used to couple amine functional groups on the ECM nanofibril surface to carboxyl moieties on the hydrogel surface. The image on the right hand side shows the hydrogel during transfer. It is pressed onto the silicon microarray between two objective slides, which are fixed with paper clamps.

5.2 Cell Adhesion to Fn and LM on Polyethyleneglycol Hydrogels

Human Foreskin Fibroblast (HFF) cells were cultured in Dulbecco's Modified Eagle's Medium (DMEM) (Invitrogen, #41966) containing Fetal Bovine Serum (FBS) (10 % v/v, #10050, Invitrogen). SH-SY5Y Neuroblastoma cells were cultured in DMEM containing FBS (20 % v/v) and penicillin/streptomycin (1 % v/v, #15140, Invitrogen). Cells were kept at 37°C and 5 % CO_2 and used for experiments below passage number 10. Following trypsinization, 10^5 cells were plated into a 35 mm petri dish containing the hydrogel.

5.3 Immunostaining of ECM Proteins

After 4h, cells were fixed for 30 minutes in pre-warmed PBS containing PFA (3.5 % w/v, #30525-89-4, Polysciences Inc.). Samples were subsequently incubated in blocking buffer: PBS containing BSA (1 % w/v) and Triton-X (0.1 % v/v). Primary antibodies against LM (#L9393, SIGMA-Aldrich), COL I (#C2456, SIGMA-Aldrich) and Fn (#F0731, SIGMA-Aldrich and #AB2033, Chemicon) were diluted 1:2000 in blocking buffer and incubated over night at room temperature. Samples were subsequently rinsed with PBS and incubated in blocking buffer. Secondary antibodies (#A11001 and #A21245, Invitrogen,) were used as described above. After 1 h, samples were rinsed with PBS, fixed in PBS containing Paraformaldehyde (PFA) (3.5 %) and mounted in PBS between two circular glass coverslips. Fluorescence images were recorded on an inverted Zeiss AxioObserver microscope connected to an UltraView Spinning Disk Confocal System (Perkin Elmer) using a 63x/1.3 NA oil immersion objective.

5.4 Staining of Focal Adhesions on LM and Fn nanofibrils

Substrate preparation and cell adhesion to PEG hydrogels was performed as described above, but with a modified fixation step: after 2 minute pre-fixation in 3.5 % PFA solution containing 0.5 % v/v Triton-X-100, cells were fixed for further 45 minutes in 3.5 % PFA solution. During the second immunostaining incubation, a mouse anti-paxillin antibody was used (#610051, BD Biosciences), which was pre-labelled with a zenon anti-mouse labelling kit (#Z25060, Invitrogen).

Chapter

6

FRET Analysis of Fn

In order to investigate the degree of unfolding of Fn molecules within Fn nanofibrils, a labeling scheme identical to the one used by Little et al. [11] was used (Figure 6.1).

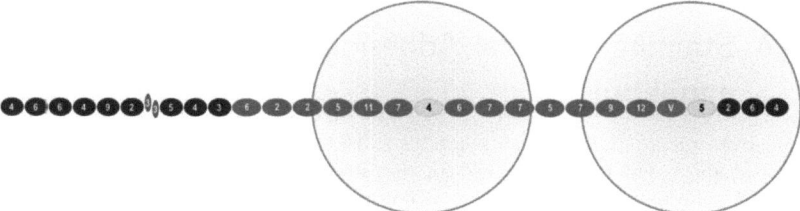

Figure 6.1: FRET Labeling Scheme of Fn. Each dimer arm of Fn consists of tandem repeats of type I (dark blue ovals), II (narrow, green ellipses), and III modules (red ovals). Average end-to-end lengths of each module type are drawn to scale using lengths of 2.5 nm for FnI [40], 0.7 nm for FnII [41], and 3.2 nm for FnIII modules [42]. Free cysteines are present on modules $FnIII_7$ and $FnIII_{15}$ (yellow). Energy transfer is limited to within approximately double the Förster radius (12 nm), denoted by circles around the acceptor sites. The numbers printed within each module represent the numbers of potential donor labeling sites.

6.1 FRET labeling of Fn

6.1.1 Thiol-reactive Acceptor Labeling

ATTO 550 maleimide (#AD 550-41, ATTO-Tec) was dissolved to 2 mM in DMSO. Fn in 6 M urea was thawed and mixed with an equal volume of PBS containing 8 M GdnHCl, resulting in a final volume of 2 ml and a Fn concentration of 1 mg ml^{-1}. To this solution, 30 µl of the ATTO 550 maleimide stock solution was added and incubated in the dark for 2 h on a rotisserie at room temperature. Unreacted dye was separated from protein by size exclusion chromatography using a PD-10 column (#17-0851-01, GE Healthcare) according to the manufacturer's protocol. As an elution buffer the reaction buffer of the subsequent labeling reaction of amino groups was used, which consisted of PBS containing 0.1 M $NaHCO_3$ at pH 8.7.

6.1.2 Amine-reactive Donor Labeling

ATTO 488 NHS-Ester (#488-31, ATTO-Tec, Germany) was dissolved to 5 mM in DMSO. To the eluate of the acceptor labeling step above, 30 µl of the ATTO 488 maleimide stock solution was added and incubated in the dark for 1 h on a rotisserie at room temperature. Unreacted dye was separated from protein by size exclusion chromatography using a PD-10 column (#17-0851-01, GE Healthcare) according to the manufacturer's protocol. As an elution buffer, PBS containing 10 % v/v glycerol was used.

6.2 Western Blotting of Labeled Fn

In order to test if fragmentation of Fn occurred during the labeling reaction, SDS PAGE and Western Blot analysis were carried out. The concentration of labeled protein fractions was estimated via absorption of the solution at 280 nm and an amount equal to 0.4 µg per lane was loaded onto the gel. Electrophoretic separation and western blotting under reducing conditions was performed using 4-20 % polyacrylamide gradient gels according to the manufacturer's protocol (#25244, Pierce). After staining in water containing acetic acid (5 % v/v) and Ponceau S (0.1 % v/v), nitrocellulose membranes were photographed and de-stained in UltraPure water. Samples with sufficiently high protein concentration were pooled and the protein concentration as well as the degree of labeling were estimated as described below.

6.3 Determination of Degree of Labeling

Absorption measurements were performed on a spectrophotometer (Nanodrop 1000, peqlab, Germany). To determine the protein concentration, correction factors for dye absorption were used as follows:

$$c_{ATTO550} = \frac{OD_{550nm}}{\epsilon_{ATTO550,550nm}}$$

$$c_{ATTO488} = \frac{OD_{501nm} - f_{ATTO550,501nm} OD_{550nm}}{\epsilon_{ATTO488,501nm}},$$

$$OD_{280nm,ATTO488} = f_{ATTO488,280nm} c_{ATTO488} \epsilon_{ATTO488,501nm},$$

$$OD_{280nm,ATTO550} = f_{ATTO550,280nm} OD_{550nm},$$

$$c_{FN} = \frac{OD_{280nm} - OD_{280nm,ATTO550} - OD_{280nm,ATTO488}}{\epsilon_{Fn,280nm}},$$

with the following values derived from the manufacturer's data sheet:

$\epsilon_{ATTO550,550nm} = 1.2 \times 10^5 M^{-1} cm^{-1}$,
$\epsilon_{ATTO488,501nm} = 9 \times 10^4 M^{-1} cm^{-1}$,
$\epsilon_{Fn,280nm} = 3.64 \times 10^5 M^{-1} cm^{-1}$,
$f_{ATTO550,501nm} = 0.172$,
$f_{ATTO488,280nm} = 0.1$,
$f_{ATTO550,280nm} = 0.12$.

6.4 FRET Calibration to Molecular Unfolding of Fn

In order to use FRET labeled Fn as an indicator of molecular unfolding, a calibration between the degree of unfolding and the FRET signal was necessary.

6.4.1 Circular Dichroism (CD) Spectroscopy

Both FRET labeled Fn and unlabeled Fn were examined using the 229 nm absorption band of Fn. CD spectra in the range 220-260 nm were recorded using a quartz cuvette (#105.250-QS, Hellma, Germany) on a Jasco model J-715 CD Spectropolarimeter (JASCO, Germany) with a step size of 1 nm, slit widht of 5 nm and integration time of 1 s. For each protein and GdnHCl concentration measured, 3 independent samples were analyzed:

Per sample, 15 µg of protein were diluted to a volume of 150 µl in PBS. Step-wise addition of 25, 50 and 75 µl GdnHCl resulted in a dilution series of protein with buffers containing 1 M, 2 M and 4 M GdnHCl, respectively.

Baseline subtraction of spectra was performed by subtracting the buffer solution signal. Instrument drift was taken into account by subtracting the mean value of the signal in the interval 245-250 nm, where a CD signal of zero is expected [66]. Pipetting errors were accounted for by applying a scaling factor to the curves. For each GdnHCl concentration, this value corresponded to the mean value of the ratios in the spectral region 224-234 nm between each spectrum and an arbitrarily chosen reference spectrum. The mean and standard deviation of the CD spectrum were converted into Mean Residue Ellipticity (MRE) values according to the following formula:

$$\mathrm{MRE}(\Theta_{229nm}) = \frac{\Theta_{229nm}\mathrm{MRW}}{lc}, \tag{6.1}$$

with the measured ellipticity, $[\Theta]_{229nm}$ in millidegree (mdeg), a Mean Residue Weight (MRW) for Fn of 108 g mol^{-1}, a measurement path length, l, of 1 cm and the protein concentration c, in g ml^{-1}.

6.4.2 FRET Signal During Unfolding of Fn

Native Polyacrylamide Gelelectrophoresis (PAGE) of Fn

Gel and buffer compositions are given in Tables 6.1 and 6.2. An amount corresponding to 0.3 µg of FRET labeled Fn was loaded per lane and the gel was run on ice for 2 h at a limiting voltage of 200 V on a BioRad mini protean system. Longer running times resulted in further migraiton of the protein band without impact on the calibration result.

Imaging of Immobilized FRET Probes

After running the native PAGE, individual bands were cut out and glued into 60 mm petri dishes using tissue glue (Nexaband, Abbot). These bands were immersed in PBS and subjected to varying concentrations of GdnHCl. Samples were imaged on a Leica SP5 Confocal Laser Scanning Microscope (CLSM) using a 20 x / 1.0 NA water dipping objective. Relevant imaging paramters are depicted in Table 6.3. Data analysis was performed as described in Section 6.6.2, except that the background signal was determined by imaging a gel which only contained PBS and no FRET probe.

6.5 FRET Measurements of Fn Surface Films

One milliliter of PBS was deposited on top of a superhydrophobic PDMS micropillar array. Unlabeled (98.75 %) and FRET-labeled (1.25 %) Fn was added to a yield a final concentration of 50 µg ml^{-1}. After 2 hours, the substrate was immersed in PBS. Donor and acceptor images of three randomly chosen areas were taken using the parameters depicted in Table 6.3. Background correction was achieved by imaging the air-buffer interface at positions on the micropillar array where no protein adsorption took place.

Table 6.1: Gel composition for native PAGE of Fn.

Reagent Name	Stacking Gel (5 %)	Separating Gel (10 %)
UltraPure water [ml]	3.4	7.9
30 % Acrylamide solution [ml]	0.83	6.7
Tris (1 M, pH 6.8) [ml]	0.63	-
Tris (1.5 M, pH 8.8) [ml]	-	5
Ammonium Persulfate (APS)[µl]	50	200
Tetramethylethylenediamin (TEMED)[µl]	5	20

Table 6.2: Buffer composition for native PAGE of Fn.

Reagent Name	Amount
Tris (Base)	30.3 g
Glycine	144 g
UltraPure water	ad 1 l

Table 6.3: Imaging parameters used for FRET calibration and imaging of Fn Nanofibrils. After performing a stability test of both available lasers, the Argon ion laser was chosen due to significantly lower power fluctuations, as shown in Figure A.2.

Parameter	Value
Scanning Frequency	100 Hz
Pinhole Size	2 Airy Units (AUs)
Laser Type	Argon-Ion
Excitation Wavelength	488 nm
Laser Output	0 % (Standby)
Laser Accusto Optical Tunable Filter (AOTF) Power	< 40 %
Bandwidth Detector 1	510 - 540 nm
Bandwidth Detector 2	555 - 595 nm
PMT Voltage Detector 1 & 2	650 V

6.6 FRET Measurements of Fn Nanofibrils

6.6.1 FRET Imaging of Fn Nanofibrils

After nanofibril depsition on stretchable PDMS microarrays, the substrate was wetted in PBS containing MOWIOL-488 (0.1 % w/v). On 3 independent samples, donor and acceptor images of three randomly chosen areas were taken using the parameters depicted in Table 6.3.

6.6.2 Data analysis

For each image, the background signal was determined by selecting a region between micropillars, where no fluorescence signal was detected. After smoothing the image with a 5 x 5 gaussian filter of radius 1 pixel, the mean background intensity was subtracted for both the donor and acceptor channel individually. The resulting pixel-wise ratio between donor and acceptor image was used as a FRET index. For each pixel $P(x,y)$ with acceptor signal $S_{Acceptor}(x,y)$ and donor signal $S_{Donor}(x,y)$, the FRET Index, $I_{FRET}(x,y)$, was thus:

$$I_{FRET}(x,y) = \frac{S_{Acceptor}(x,y) - BG_{Acceptor}}{S_{Donor}(x,y) - BG_{Donor}} \qquad (6.2)$$

To minimize noise, low intensity pixels were excluded by applying a dynamic threshold of five standard deviations of the background signal.

Chapter

7

Mechanical Properties of Fn Nanofibrils

7.1 Extensibility of Fn Nanofibrils

Nanofibrils produced on stretchable silicone substrates were subjected to step-wise extension from 100 % to 300 % elongation. The overall process of stretching endured for one hour and was imaged on a Leica SP5 CLSM system using a 20 x / 1.0 NA Water Dipping Objective.

7.2 AFM of Fn nanofibrils

7.2.1 Sample Preparation

Immediately after production of Fn Nanofibrils on PU substrates, the glass coverslip was glued into a 60 mm petri dish, which had a hole cut in the middle (see Fig 7.1). Since the PU micropillar array wetted spontaneously after several minutes, there was no need to use a surface active substance such as MOWIOL-488. Complete wetting of the substrate was achieved after 1 h of incubation at room temperature.

The samples were mounted on a Leica DMI 6000 CS inverted microscope equipped with a 40 x / 0.65 NA objective. A JPK Nanowizard II scan head was operated via the provided SPM Software Release 3.3. Fibrils were selected manually and position information was recorded using the Micro Motion nanopositioning stage provided with the instrument. This position information was later used for the correlation of single nanofibrils imaged in Scanning Electron Microscopy (SEM) with the corresponding AFM force curves.

Force-displacement curves were acquired in 1 μm steps along each selected nanofibril using a tipless cantilever (CSC12, MikroMash, Estonia), which was calibrated according to the SPM Software manual prior to the experiment. The parameters used for force curve acquisition are depicted in Table 7.1.

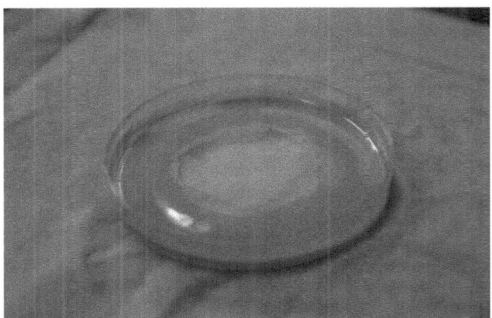

Figure 7.1: AFM Sample. The microstructure is fixed on a circular glass coverslip which is glued into the lid of a petri dish. This assembly can be filled with buffer and placed under the scan head of a JPK Nanowizard AFM.

Table 7.1: Parameters used for AFM force imaging of Fn Nanofibrils.

Parameter	Value	Description
Force	8 nN	Threshold force
Force setting	relative	Relative force measurement
Duration	0.5 s	Approach duration
Z length	5 µm	Length of force scan
Feedback Mode	Contact	AFM imaging mode used

7.3 SEM of Probed Fn nanofibrils

After finishing the AFM experiment, the sample was recovered and fixed in PBS containing 3.5 % PFA for 1 h at room temperature. After copious rinsing in UltraPure water, the sample was left to dry over night at room temperature.

The next morning, the samples were recovered and sputter coated with 2 nm Gold (0.05 mbar, 30 mA, 10 s) in a BalTec MCS 010 sputter coater to reduce charging effects of the polymer during imaging.

Using fluorescence images acquired during AFM imaging combined with position data derived from the nanopositioning stage of the AFM, it was possible to identify single nanofibrils which were probed during the AFM experiment. The diameter of these fibrils was determined at three positions along the fibril and the average value was used in the subsequent data analysis.

7.4 Data Analysis of AFM Experiments

The Data analysis was based on a three-point bend test [57]. A custom-written MATLAB script (The MathWorks, Inc., Natick, Massachusetts) was used to convert and analyze the force curves derived from several AFM experiments.

Part III

Results and Discussion

Chapter

8

Surface Activity of ECM Proteins

While the effect of plasma molecules such as Bovine Serum Albumin (BSA) [67] and immunoglobulins [68] on surface tension has been intensively studied, little is known about the surface activity of ECM proteins. Protein accumulation on superhydrophobic surfaces followed by surface de-wetting is required for nanofibril formation. Thus, the investigation of the surface activity of ECM proteins is essential for a detailed understanding of the underlying mechanisms.

In addition to their tendency to accumulate at the air-buffer interface, the self-assembly qualities of ECM proteins were underlined by their ability to form fibrils when pulled out of a surface film, which formed spontaneously (Figure 8.1). Contrasting this behavior, BSA accumulates at the air-buffer interface but does not form fibrils.

8.1 Protein Accumulation at the Air-Buffer Interface

The dynamics of protein accumulation at the air-buffer interface were investigated on a Langmuir film balance [69] over a time span of 1 hour. For all proteins, a fast increase in surface pressure was observed within 30 seconds after protein addition to the subphase, as shown in Figure 8.2.

8.2 Discussion

8.2.1 Dynamics of Protein Accumulation at the Air-Buffer Interface

The accumulation of ECM proteins at the air-buffer interface is a rapid process, which takes place on the time scale of 30 s. The same characteristics have been described for BSA in previous studies [67, 70].

In these earlier investigations, it could also be shown that the increase in surface pressure caused by protein accumulation at the air-buffer interface is largely independent of protein concentration, with most of the changes occurring within the

Figure 8.1: Pulling Fibrils from Surface Films of ECM Proteins. Top: After addition of ECM protein to a layer of PBS, an equilibrium forms between molecules in solution and at the air-buffer interface. The film can be compressed and fibrils can be manually pulled using a pair of tweezers. Bottom: The image shows a fibril consisting of COL I.

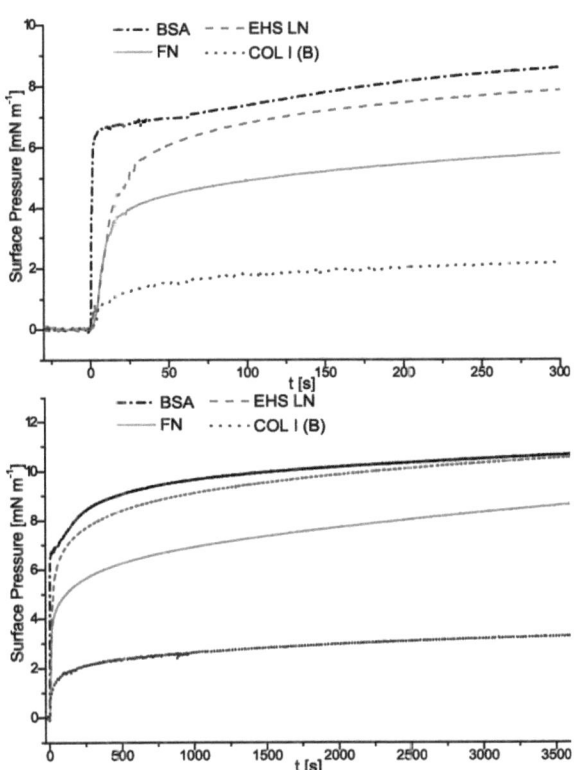

Figure 8.2: Surface Activity of Extra Cellular Matrix (ECM) Proteins and BSA. Top: Already after 120 seconds the surface pressure rises to more than 75% of the value reached after 1 hour (bottom). Addition of BSA (dash-dotted line, black) results in a rapid increase in surface pressure. The surface pressure resulting from LM (dashed line, red) was the highest of all three ECM proteins.

first hour of observation. Information derived from these experiments can thus be applied to the nanofibril fabrication method described herein.

A striking feature observed for the protein surface films reported in literature is the fact that the accumulated molecules are immobilized at the interface [71]. For Fn, this could be demonstrated by Ulmer et al. [72], as shown in Figure 8.3. The same behavior was observed when photobleaching acceptor dye on FRET-labeled Fn within surface films, as shown in Figure 10.7.

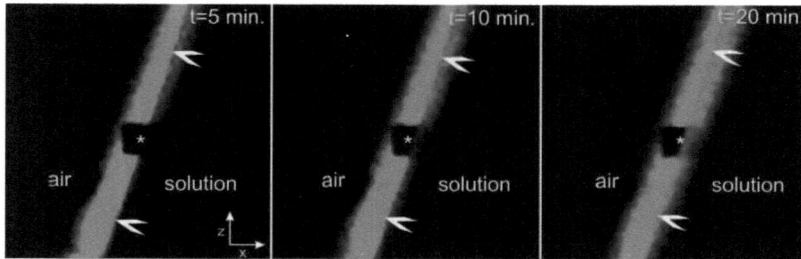

Figure 8.3: Formation of an Immobile Fn Layer at the Air-Buffer Interface. The images show a time series of confocal fluorescence micrographs, displaying the protein exchange dynamics at the air-buffer interface after photobleaching. Dye bleaching was initiated 1 h after injecting Fn into a PBS droplet, and was followed by injecting fresh Fn. The original protein border, which formed after 1 h, is indicated by the white arrowheads. Fluorescence recovery after photobleaching was measured in the area denoted with an asterisk. While fresh Fn readily interacted with the bleached zone, the original interface essentially did not recover during the 20 min timeframe of this experiment. Taken from [72].

8.2.2 Structural Consequences of Protein Accumulation

Protein accumulation and structural changes at air-buffer interfaces have been studied for several technologically relevant molecules so far, ranging from BSA [73] over immunoglobulins to enzymes such as horseradish peroxidase [74] and lysozyme [75]. The emerging picture is that at least part of the molecules undergo gradual loss of tertiary and secondary structure when forming an immobilized film of molecules at the air-buffer interface. In this context, work by Vogel et al. [76] and Garcia et al. [77] has demonstrated that structural changes of Fn occur upon its adsorption to interfaces and influences differentiation and proliferation of myoblast cells [77].

A second potential effect of protein accumulation at the air-buffer interface could be the formation of intermolecular sulfhydryl crosslinks, which was demonstrated for ovalbumin by Lechevalier et al. [75].

This leads to the suggestion that at least parts of the ECM molecules might become unfolded upon accumulation at the air-buffer interface. Particularly, the non-disulfide bonded type III modules of Fn might become unfolded, exposing cryptic self-binding sites (see Figure 1.2), which might result in the formation of intermolecular disulfide bridges. Although the existence of free sulfhydryls on modules FnIII$_7$ and FnIII$_{15}$ could allow such an interaction and work by Patel *et al.* recently demonstrated its occurrence *in vitro* [78], it has to be noted that the existence of intermolecular disulfide bonds in Fn fibrils has been questioned [79].

An investigation of the secondary structure of all ECM proteins immobilized at the air-buffer interface was beyond the scope of this work, but could be achieved for Fn by using the FRET technique, as described in Section 10.3.

Chapter

9

Size Control of ECM Nanofibrils

The goal of this work was to control the alignment, diameter and tensile state of Fn nanofibrils. Recent work in our lab demonstrated the possibility of Fn fibrillogenesis by wetting-induced surface forces on PDMS micropillars [72]. However, this method enabled limited control over fibril directionality and nanofibril diameter, which both influence cellular response to these surfaces.

A de-wetting based method was thus developed which allows control over directionality and diameter of protein nanofibrils. Initial experiments showed that Fn formed fibrils when manually pulling a drop of Fn along a PDMS micropillar structure. Subsequently, the parameters pulling speed, v, protein concentration, c, and array geometry were varied in order to investigate their influence on nanofibril diameter. The picture that emerged from experiments on silicon micropillar arrays is that array geometry predetermines the distribution of Fn nanofibril diameters, while pulling speed and protein concentration have to be optimized for each geometry to achieve full coverage and narrowest distributions of nanofibril diameters.

As an extension of the work presented here, this method allowed the production of EHS LM and COL I nanofibrils as well as an application to moulded PDMS or PU micropillar structures.

9.1 Silicon Microarrays Produced by Reactive Ion Etching (RIE)

The de-wetting behavior of the surfaces used herein, and thus the quality of the produced nanofibrillar arrays depends on the regularity of the microstructure used. A method to produce durable silicon micropillar substrates was established, which allowed repeated cleaning and reuse of these sensitive substrates.

Figure 9.1: Silicon Masks. Depending on the number of etching cycles, n, the height of the micropillars can be varied. Left: n = 100; middle: n = 150; right: n = 200. Small debris attached to the structures during cutting of the silicon wafer. Scale bar: 5 µm.

9.2 Fn Nanofibrils on Micropillar Substrates

During initial experiments, a 10 µl drop of Fn in PBS was manually pulled over a superhydrophobic micropillar surface. While the resulting Fn fibrils aligned parallel with the pulling direction in the central region of the meniscus, they were randomly oriented at lateral areas (see Figure 9.2), where the receding meniscus was highly curved.

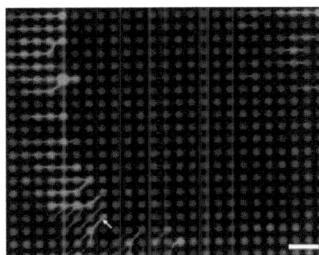

Figure 9.2: Experiment to Produce Fn Nanofibrils. Fibronectin fibrils formed when manually pulling a 10 µl drop of Fn from left to right over a PDMS micropillar array. The fibrils at the curved area of the moving meniscus point towards the center of the droplet (white arrow). Scale bar: 50 µm.

The control of nanofibril directionality was achieved by moving the substrate relative to a droplet that is spread along a glass rod, as shown in Figure 9.3. This resulted in a meniscus that is perpendicular to the pulling direction along the micropillar array.

Using hydrophobic silicon micropillar arrays, the optimum of both the solution concentration of Fn and the pulling speed were determined for each substrate geometry. Subsequent SEM analysis on silicon micropillar arrays provided the distribution of fibril diameters for each substrate geometry. The parameters depicted in Figure 9.4 yielded optimal regularity and narrowest distributions of Fn nanofibril diameters over substrate areas as large as 2 cm^2.

The same parameters could be equally applied on PDMS micropillar arrays, which have the advantage of being both transparent and elastomeric and can be used in force-sensing experiments (see Figure 9.5) [80].

The observation that lower protein concentrations are needed to produce nanofibrils on small micropillars (d = 2.5 µm, see Figure 9.4) proved useful in the investigation of other protein systems. Thus, all further experiments employing ECM protein

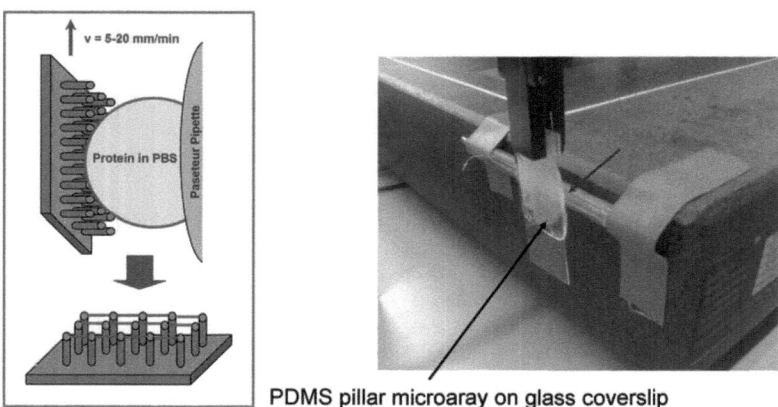

Figure 9.3: Fabrication of ECM Fibrillar Arrays. The image on the right hand side shows the setup which was used to create the fibrillar arrays. The protein solution was applied on the accessible glass surface of the pasteur pipet (blue arrow).

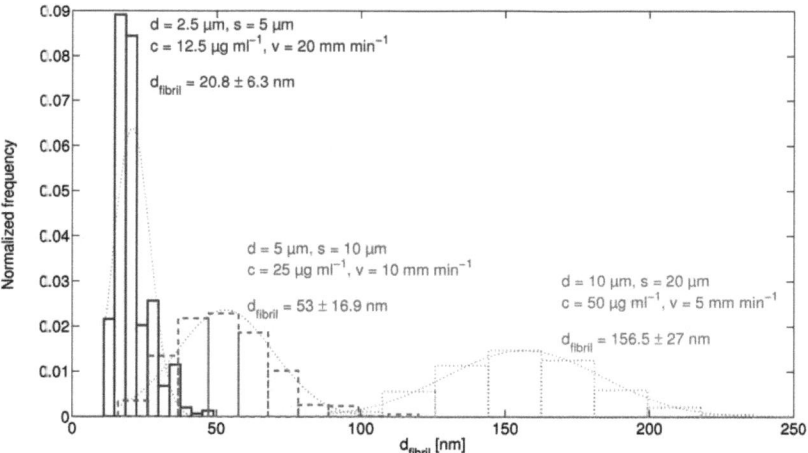

Figure 9.4: Diameter Control of Fn Nanofibrils. Histograms and fitted normal distributions of Fn nanofibril diameter for the microarray geometries depicted in Figure 9.5. The inset text indicates the parameters pillar spacing, s, pillar diameter, d, (see Figure 3.1), protein concentration, c, and pulling speed, v. For each histogram, 350 fibrils over three independent experiments were analyzed.

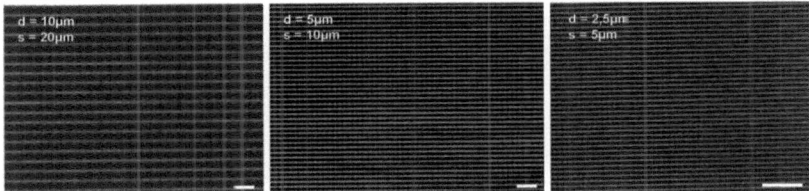

Figure 9.5: Regular Arrays of Fn Nanofibrils. Fluorescence micrographs of ATTO 488 labeled Fn fibrils produced on PDMS microarrays with different geometries. The pillar diameter, d, and center-to-center separation, s, define the microarray (see Figure 3.1). Protein concentration and pulling speed correspond to the values given in Figure 9.4. Scale bar: 40 µm.

components were first performed on arrays with the smallest micropillar diameters available.

9.2.1 Discussion

Control of Nanofibril Directionality

Fn nanofibrils align perpendicular to the meniscus of a protein solution. This indicates that the surface forces acting during de-wetting of the superhydrophobic microarray play a central role in nanofibril formation. Guan et al. proposed a model of nanofibril formation for a similar system, as shown in Figure 9.6. The observations made during this work support this model.

Control of Nanofibril Diameter and Array Regularity

Similar to findings by Guan et al. [82], who used an analogous method to fabricate nanowires consisting of Deoxyribonucleic Acid (DNA) 82, 81] and synthetic polymer [83], the diameter of Fn nanofibrils mainly depends on the diameter of the underlying pillars. In contrast to the fabrication of DNA nanowires by Guan et al., the range of Fn concentrations and pulling speeds at which regular Fn nanofibril arrays could be accomplished was specific for each substrate geometry. Applying too high concentrations of protein or too low pulling speeds resulted in the formation of thick and randomly oriented fibrils, while low Fn concentrations or higher pulling speeds led to defects in the nanofibrillar array (see Figure 9.7).

One explanation for this could be the tendency of Fn to accumulate at the air-buffer interface [72] (see Figure 8.2). This property is not observed for DNA [73] and introduces the dynamics of protein diffusion and accumulation at the air-buffer interface to the fabrication process.

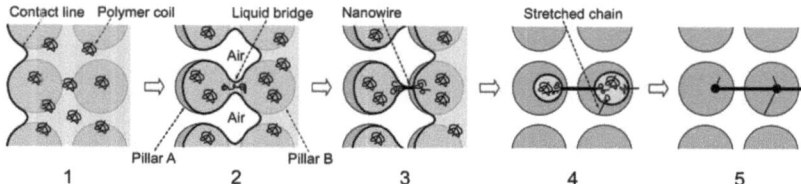

Figure 9.6: Model of Nanofibril Formation. The aqueous polymer solution moves from left to right with the thick lines designating the water fronts or contact lines. After the contact line passes the narrowest point of the gap between adjacent pillars in the vertical direction, the contact line jumps to the next pillar row, as shown in 2. A "liquid bridge" forms between the two adjacent pillars A and B. The liquid bridge is prone to rupture because of its high surface-to-volume ratio and high surface tension of the aqueous polymer solutions. Water evaporation may also contribute to this instability. The liquid bridge breaks in step 3, creating two separated liquid bodies. A single polymer molecule at the breaking point can be trapped in both liquid bodies and stretched in the opposite directions. Several of these stretched molecules are bundled together by capillary forces to form a nanowire. In step 4, the isolated droplet on pillar A spontaneously shrinks owing to the hydrophobic nature of PDMS, continually extending the nanowire leftward as a result of molecular combing. At the same time, a droplet forms on pillar B. This droplet also beads up, leading to the growth of the nanowire in the right direction. On the other side of the droplet on pillar B, another nanowire forms and grows leftward, repeating the process from pillar A. The droplet eventually dries and a dotlike precipitate forms at a point where the two nanowires meet, as shown in step 5. This process repeats itself on all pillars when the water front passes, finally generating a long, continuous nanowire on the pillars. Modified from [81].

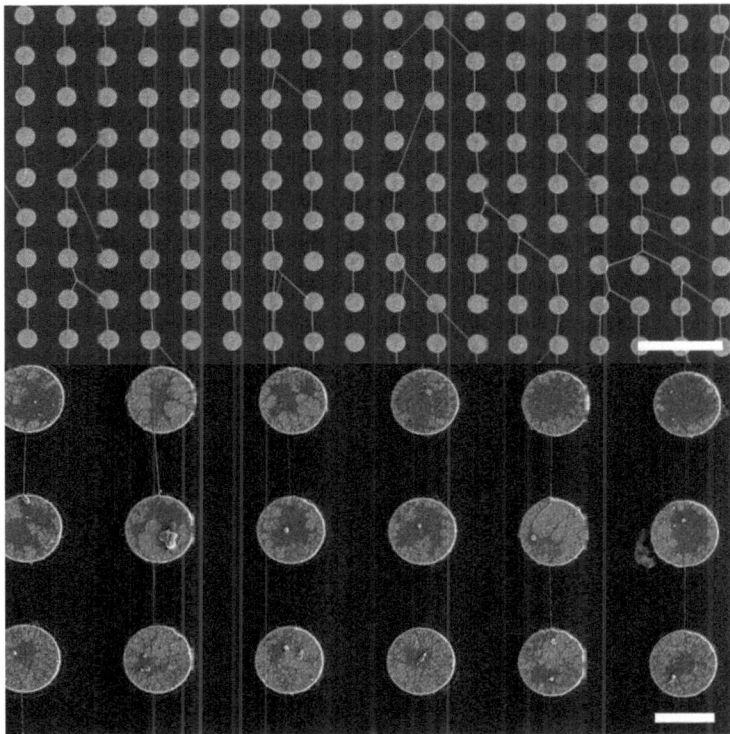

Figure 9.7: SEM Images Illustrating Concentration and Pulling Speed Dependence of Fn Nanofibrils. Top: An increase in Fn protein concentration and reduction of pulling speed leads to a higher amount of irregular fibrils on arrays with 2.5 µm pillar diameter. Bottom: Likewise, using 2-fold lower than optimal protein concentrations and 2-fold increased pulling speeds, Fn nanofibrillar arrays show defects on arrays with 10 µm pillar diameter. Fn was used at a concentration of 25 µg ml^{-1} (top) and 12.5 µg ml^{-1} (bottom), respectively. In both experiments, the pulling speed was 10 mm min^{-1}. Scale bar: 10 µm.

Self-Association Sites are Necessary for Nanofibril Formation

Different from DNA molecules, which are negatively charged and thus self-repellent, Fn has several self-binding sites [18], which favors the formation of cross-connected nanofibrils, as shown in Figure 9.7. A necessity for accessible self-binding sites is supported by recent work by Pabba *et al.* [84], where fibrinogen formed nanofibrils only in the presence of its activator thrombin.

We could further prove the requirement of self-binding sites by testing nanofibril formation of BSA on silicon micropillar arrays, as shown in Figure 9.8. In line with these observations, BSA did not form nanofibrils at high protein concentrations of up to 1 mg ml^{-1}, indicating that proteins which do not form fibrils *in vivo* will not form fibrils using the described method.

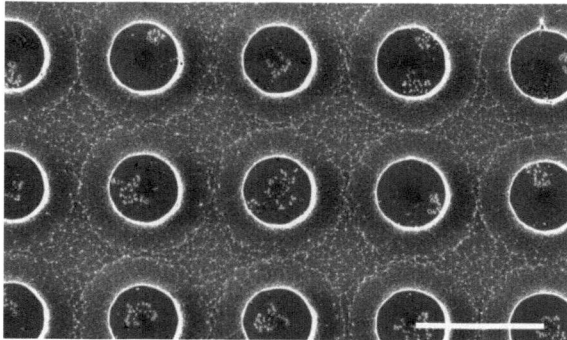

Figure 9.8: Nanofibril Formation Requires Self-Binding Sites. BSA, even when employed at concentrations of 1 mg ml^{-1}, does not yield nanofibrils at a pulling speed of 2 mm min^{-1}. Scale bar: 5 μm.

9.3 Fabrication of LM and COL Nanofibrils

All ECM proteins investigated here accumulate at the air-buffer interface and are known to have self-binding or even self-assembly properites (see Section 2. The hypothesis that the proteins under investigation also form nanofibrils using the method described herein was thus tested.

9.3.1 LM Nanofibrils

Laminin (LM) purified from Engelbreth-Holm-Swarm tumors (LM-111) yielded regular arrays of nanofibrils at 4-fold higher protein concentrations and 10-fold lower

Fabrication of LM and COL Nanofibrils

pulling speeds (Figure 9.9) compared to Fn. LM accumulates rapidly at the air-buffer interface, as shown in Figure 8.2). Compared to Fn, the LM molecule contains fewer self-binding sites (Section 2, which explains the lower tendency to form nanofibrils.

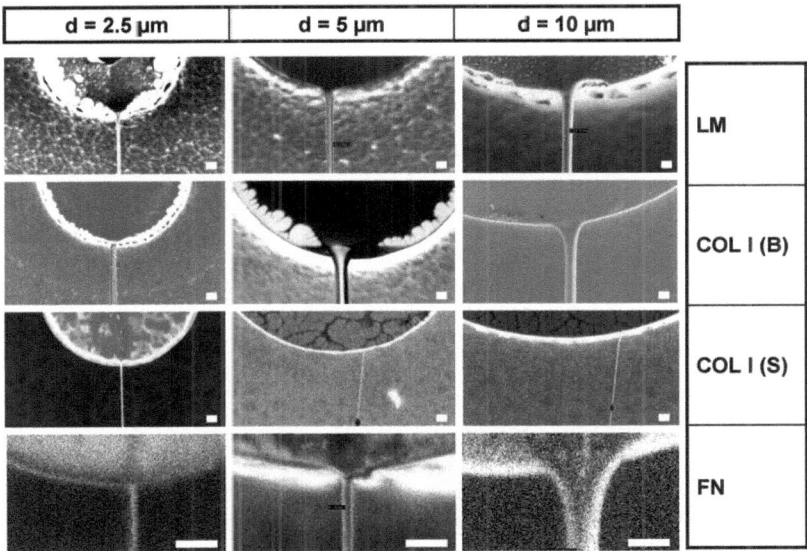

Figure 9.9: SEM Images of Nanofibrils Formed by Different ECM Proteins. Like Fn, both LM (EHS LM) and COL I (B) formed nanofibrils with increasing thickness upon increasing the micropillar diameter. COL I (S) and COL I (C) (not shown) consistently yielded nanofibrils with diameters below 30 nm, independent of array geometry, pulling speed and protein concentration. Scale bar: 200 nm.

9.3.2 COL I Nanofibrils

Initial experiments using COL I produced highly regular arrays of nanofibrils with diameters below 30 nm independently of pillar diameter, protein concentration and pulling speed. Three collagen preparations from two different providers were subsequently tested for their ability to form nanofibrils (Figure 9.9):

1. COL I purified from calf skin (SIGMA-Aldrich), referred to as COL I (C)
2. COL I purified from rat tendon (SIGMA-Aldrich), referred to as COL I (S)
3. COL I purified from rat tendon (Harbor BioProducts), referred to as COL I (B)

In summary, both COL I (S) and COL I (C) yielded fibrils with diameters below 30 nm at concentrations up to 0.5 mg ml^{-1}. COL I (B) yielded fibrils with diameters above 50 nm at concentrations below 0.2 mg ml^{-1}. All stated protein concentrations are based on information provided by the manufacturer.

9.3.3 Western Blot Analysis of ECM Preparations

Since all protein preparations are isolated from ECM of living organisms, Western Blot analysis was performed to rule out potential contamination of ECM protein preparations with Fn, as shown in Figure 9.10. Supporting this evidence, immunofluorescence staining of the produced LM fibrils was positive for EHS LM and negative for Fn, as shown in Figure 9.13. The same result was observed for COL I when immunostaining Fn, as shown in Figure A.1.

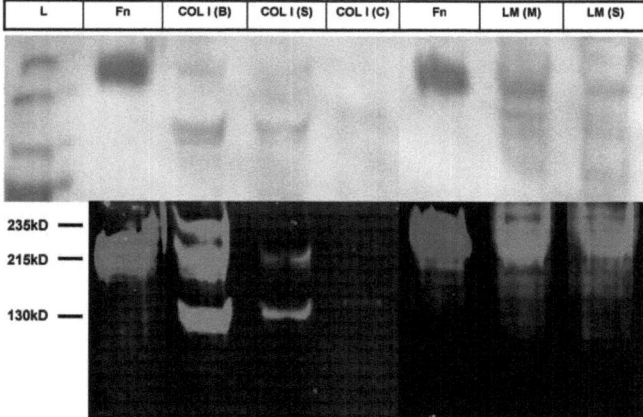

Figure 9.10: Western Blot of ECM Protein Preparations. Top: Ponceau S stain containing the molecular weight ladder, L. This stain indicates the presence of protein on the nitrocellulose membrane before immunoblotting. Bottom: COL I (S) showed reduced amounts COL I alpha chains. No cross-contamination of either LM or COL I preparations with Fn (red) could be detected using an immunofluorescence assay. Two different LM preparations from different manufacturers (M - Millipore, S - SIGMA) showed identical characteristics. Each lane was loaded with 2.67 µg of protein according to the manufacturer's data sheet.

9.3.4 Nanofibrils Consisting of Other Biopolymers

As an extension of this work, it could be shown that both actin and LM-511 form nanofibrils using the method described herein, as shown in Figure 9.11.

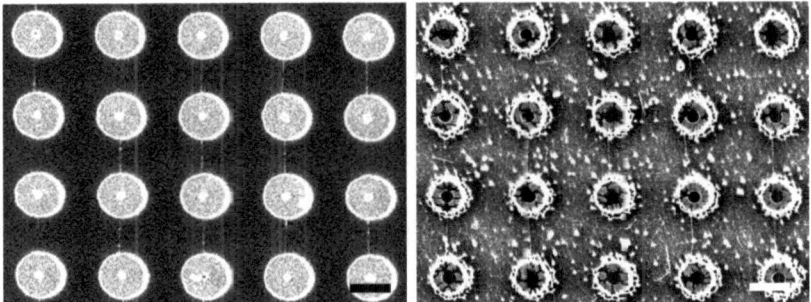

Figure 9.11: SEM Images of Actin and LM-511 Nanofibrils. Like Fn, both actin (left) and LM-511 (right) formed nanofibrils. Concentration and pulling speed was 1.3 mg ml^{-1} and 2 mm min^{-1} for actin and 0.5 mg ml^{-1} and 2 mm min^{-1} for LM-511, respectively. Scale bar: 2.5 μm.

9.3.5 Discussion

Influence of Protein Concentration and Pulling Speed on Nanofibril Formation

The formation of LM and COL I nanofibrils required higher protein concentrations and lower pulling speeds than those used for Fn. This could be due to the lower self-binding capabilities of those two proteins compared to Fn [6].

When using LM and COL solutions above a minimum protein concentration at low enough pulling speed, the formation of LM and COL I nanofibrils was rather insensitive to changes in those two paramters compared to Fn. This is in line with a reduced self-binding capability of LM and COL I molecules, respectively.

Distinct Behavior of COL I preparations

Although a transfer of the nanofibril fabrication method to other ECM molecules was possible, it was observed that the quality of ECM protein preparations largely differs between manufacturers (see Figure 9.10). This has a significant influence on the quality of the produced nanofibril arrays. A similar observation was made for commercially available Fn preparations by Ulmer [85]

The difference in detectable protein concentration could be related to the tendency of COL preparations to form nanofibrils with larger diameters. Both COL

(S) and COL (C), which are provided by the same manufacturer, show reduced protein concentration in a Western Blot experiment (Figure 9.10 and display nanofibril diameters below 30 nm independent of array geometry.

This could be either due to a significantly lower overall protein concentration or to a difference in protein composition. It is unclear at this moment, how far the level of Tropocollagen differs between all COL preparations that were used in this study. A higher relative amount of Tropocollagen could result in an increased tendency to form fibrils, even at low overall concentrations. Although the buffer composition of the stock solution was highly similar for two of the three COL preparations, it cannot be excluded that additives influence the surface tension or protein aggregation of the solutions that were diluted in PBS.

General Applicability of the Self-Assembly Method to Other Polymers

As shown in 9.11, it is possible to produce nanofibrils consisting of other biomolecules, such as actin and LM-511. The formation of polymer nanofibrils using similar methods hase been described [81, 83, 84]. This allows the creation of novel materials, either of biological or synthetic origin, for experimental and applied use.

9.4 Cell Adhesion to Fn and LM Nanofibrils

In order to demonstrate the functionality of ECM nanofibrils, a model system for cell experiments was developed, which allows the immobilization of ECM nanofibrils in any desired orientation relative to each other.

Since the ECM nanofibrils were formed by de-wetting of a superhydrophobic surface, they were in a dry state. Wetting of the micropillar surface followed by covalent attachment of nanofibrils onto PEG hydrogels reconstituted the nanofibrils in buffer and simultaneously eliminated substrate microtopography (Figure 9.12).

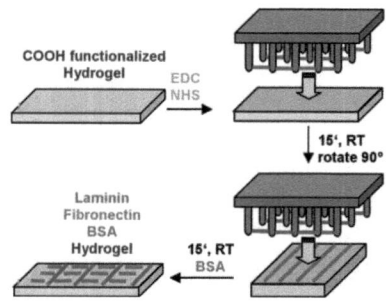

Figure 9.12: Transfer and Covalent Attachment of LM (red) and Fn (green) Nanofibrils onto PEG Hydrogels. Hydrogels were copolymerized with 2-carboxyethylacrylate to introduce -COOH functional groups. Standard EDC/NHS chemistry was used to couple amine functional groups on the ECM nanofibril surface to carboxyl moieties on the hydrogel surface.

After sequential transfer of LM and Fn nanofibrils, we investigated the differential binding of SH-SY5Y and HFF cells to LM and Fn nanofibrils, respectively (Figure 9.13). Cells adhere to the crossed ECM nanofibril network via paxillin-containing adhesion sites (Figure 9.14), which demonstrates the biofunctionality of the fibrils after their transfer. While HFF cells line rather perfectly with Fn nanofibrils and are not influenced by the directionality of LM nanofibrils, SH-SY5Y cells interact with both LM and Fn nanofibrils, but show distinct outgrowths along the LM fibrils (see Figure 9.13).

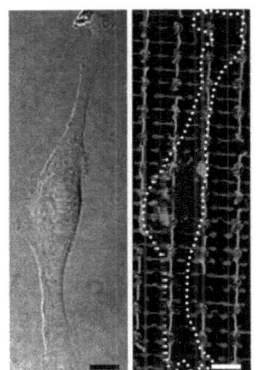
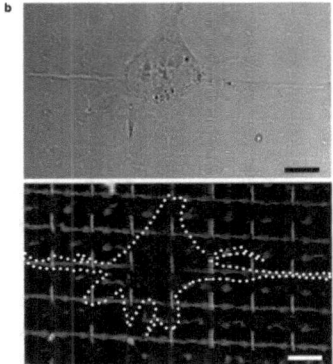

Figure 9.13: Differntial Cell Adhesion on Fn and LM Nanofibrils. (a) HFF cell adhering to Fn nanofibrils (green) on a combined LM/Fn nanofibril substrate. LM is shown in red. For reasons of visibility, the panel on the right displays the circumference of the cell seen in differential interference contrast (DIC) mode on the left panel. (b) SH-SY5Y neuroblastoma cell adhering to Fn (green) and LM (red) nanofibrils. The cell produces outgrowths along the LM fibril in the center. In addition to differential adhesive behavior, distinction between both cell types in co-culture was possible by morphological differences in size and shape. Scale bar: 10 µm. Taken from Kaiser and Spatz [36].

9.4.1 Discussion

The use of substrates with combined LM and Fn nanofibrillar arrays enables the distinction of fibroblast-like cells from neuroblastoma cells in co-culture assays due to their differential adhesion to ECM nanofibrils. Already the use of a single nanofibril array grafted on non-adhesive PEG hydrogels opens the route to cell-guiding substrates, which are similar to the ones accessible by micro-contact printing [87]. While the sequential transfer of two different nanofibrillar arrays opens potentially interesting opportunities to generate tailored substrates for selective cell adhesion, it should be noted that the fabrication process requires mechanical contact with the

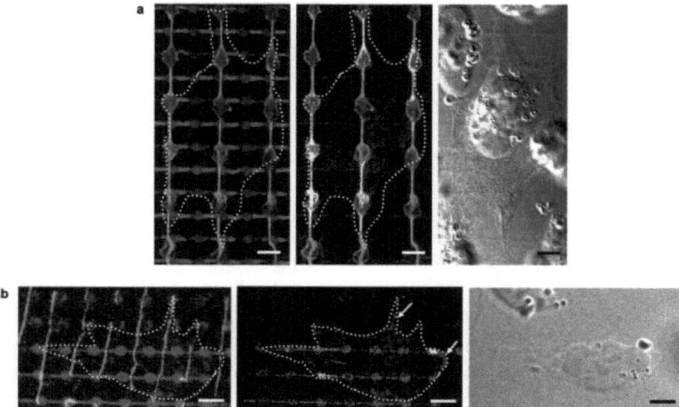

Figure 9.14: Focal Adhesion Staining on Fn and LM Nanofibrils. Colocalization of focal adhesion protein, paxillin (blue), with Fn (green) and LM (red) nanofibrils. (a) HFF cell adhering exclusively to Fn nanofibrils. White color in the center image indicates colocalization of Fn and paxillin. (b) SH SY5Y cell adhering to both Fn and LM nanofibrils. While the LM nanofibril direction dictates orientation, Fn nanofibrils support cell adhesion, as indicated by white arrows in the center image. Scale bar: 5 μm. Taken from Kaiser and Spatz [86].

nanofibril arrays, which can lead to defects in the produced patterns. A future use of these substrates on a large scale depends on the technical ability to overcome this problem.

Chapter

10

FRET of Fn Nanofibrils

The development of a stretchable array of Fn nanofibrils was possible using the methods described in Chapter 9. In order to establish a mechanically switchable Fn matrix, the possibility to alter the molecular structural properties within Fn nanofibrils was investigated using a FRET labeling approach [9, 10] (see Figure 10.1).

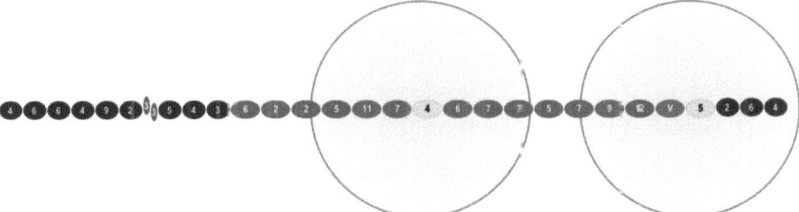

Figure 10.1: FRET Labeling Scheme of Fn. Each dimer arm of Fn consists of tandem repeats of type I (dark blue ovals), II (narrow, green ellipses), and III modules (red ovals). Average end-to-end lengths of each module type are drawn to scale using lengths of 2.5 nm for FnI [40], 0.7 nm for FnII [41], and 3.2 nm for FnIII modules [42]. Free cysteines are present on modules $FnIII_7$ and $FnIII_{15}$ (yellow). Energy transfer is limited to within approximately double the Förster radius (12 nm), denoted by circles around the acceptor sites. The numbers printed within each module represent the numbers of potential donor labeling sites. On average, each monomer was labeled with two acceptors and twenty-eight donors as determined by spectrophotometry of FRET labeled protein.

Following calibration of the FRET sensor, the unfolding state of Fn molecules at the air-buffer interface and within nanofibrils was probed. Subsequently, the FRET signals of Fn nanofibrils were compared before and after extension and rupture on a stretchable micropillar array. To rule out intermolecular FRET, a mixture of 98.75 % unlabeled and 1.25 % FRET-labeled Fn was used in all FRET experiments [10], as shown in Figure 10.2.

10.1 Calibration of Fn FRET Probe

In order to verify that FRET labeling did not alter Fn structure, the unfolding of labeled and native Fn was compared. The labeled and unlabeled protein showed identical unfolding characteristics. Following this, the FRET signal, I_{FRET}, as a function of Guanidine Hydrochloride (GdnHCl) concentration was determined for an ensemble of molecules. This provided a reference between the Fn FRET signal and the degree of molecular unfolding and allows the comparison with results from other studies [88, 10, 11].

10.1.1 Circular Dichroism (CD) Spectrometry of Fn Unfolding

The unfolding of Fn by GdnHCl was observed via the change of Fn's CD signal at a wavelength of 229 nm. This peak is attributed to ordered tyrosine and tryptophane residues within the disulfide-bonded C- and N-terminal modules of Fn [2]. The labeled FnIII modules are located at the central part of the molecule and weres shown to unfold similar to the above mentioned disulfide-bonded C- and N-terminal modules [88]. Thus, while Circular Dichroism does not directly probe the structural properties of the labeled FnIII modules, the resulting signal can be used to estimate the degree of unfolding of the whole Fn molecule.

10.1.2 Wavelength Determination for FRET Measurements

In order to determine the optimal wavelength for the FRET donor and acceptor channels, labeled Fn was immobilized via native PAGE and an emission wavelength scan was performed on the Confocal Laser Scanning Microscope (CLSM) setup used during this work. From these experiments, the optimal detection wavelength interval is [555 nm, 595 nm]. Donor emission was recorded along the interval [510 nm, 540 nm].

10.1.3 FRET as a Function of Fn Unfolding

FRET signals of Fn probes were recorded at increasing concentrations of denaturing agent (GdnHCl) in order to correlate Fn's structural properties with the FRET signal, I_{FRET}:

$$I_{FRET} = \frac{S_{Acceptor} - BG_{Acceptor}}{S_{Donor} - BG_{Donor}}, \tag{10.1}$$

with acceptor signal, $S_{Acceptor}$, donor signal, S_{Donor}, and the respective background signals, $BG_{Acceptor}$ and BG_{Donor}. The resulting calibration histograms are plotted in Figure 10.4.

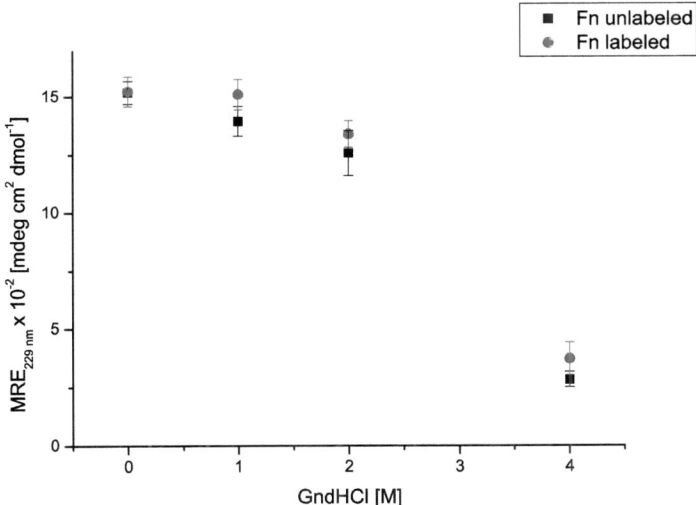

Figure 10.2: CD Analysis of Fn Unfolding With and Without FRET Labels. The error bars correspond to the standard deviation from three independent experiments. No significant structural transition occurs below a Guanidine Hydrochloride (GdnHCl) concentration of 1 M. A drastic conformational change occurs between 2 M and 4 M denaturing agent. Within measurement accuracy, no difference between both Fn preparations is found.

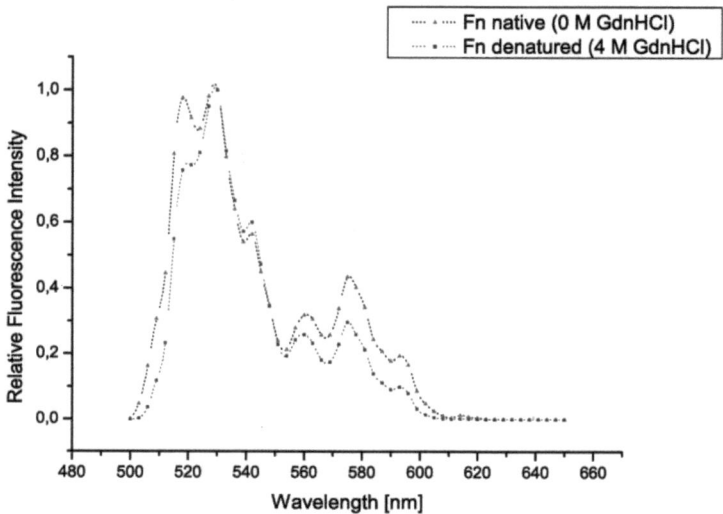

Figure 10.3: Wavelength Scan of FRET Probe. At an excitation wavelength of 488 nm, the relative emission of the FRET probe was recorded in the range between 500 nm and 650 nm both in the native (red) and denatured (blue) state. The optimal detection wavelength interval is [555 nm, 595 nm]. Donor emission was recorded along the interval [510 nm, 540 nm] in the subsequent FRET experiments.

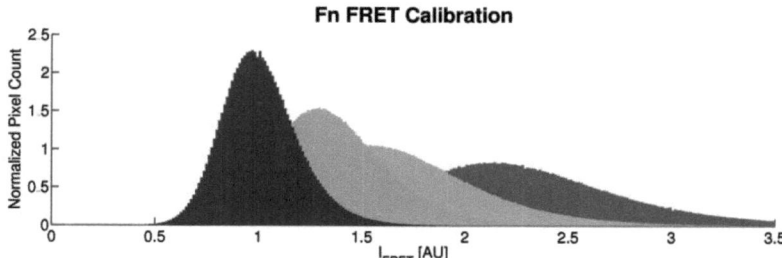

Figure 10.4: FRET Analysis of Fn Immobilized in Native Polyacrylamide Gelelectrophoresis (PAGE). Red: 0 M GdnHCl, gold: 1 M GdnHCl, cyan: 2 M GdnHCl, blue: 4 M GdnHCl. The calibration result is plotted at the bottom of the following figures in this chapter as a reference.

Calibration of Fn FRET Probe

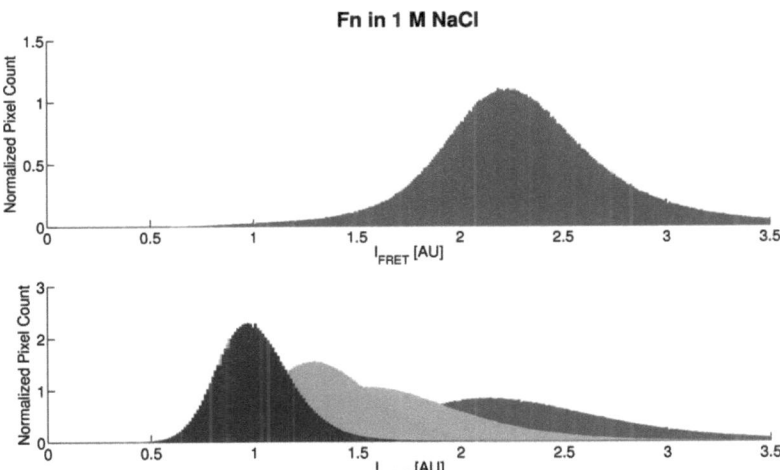

Figure 10.5: FRET Analysis of Fn Extended in High Ionic Strength Buffer. The FRET signal does not change upon extension of the molecule in 1 M NaCl. The bottom graph shows the calibration curves for increasing denaturant concentration. Red: 0 M GdnHCl, gold: 1 M GdnHCl, cyan: 2 M GdnHCl, blue: 4 M GdnHCl.

10.1.4 Discussion

The Unfolding Characteristics of FRET Labeled and Native Fn are identical

The secondary structure analysis of Fn and the FRET-labeled probe show identical results. Since the CD method is an ensemble measurement of secondary structural features and Fn shows only low occurrence of these features, the conclusion from this result is that a gradual change in secondary structure is inducible by addition of GdnHCl and the solution structure of labeled Fn shows no significant deviation from the structure of native Fn.

Fn Molecular Unfolding Can Be Observed by FRET

The FRET calibration versus GdnHCl concentrations allows the correlation of molecular unfolding with a fluorescence readout on a microscopic level.

In this context, an important distinction has to be made between extension (loss of tertiary structure) and unfolding (loss of secondary structure) of Fn molecules. The solution structure of Fn appears to be a string of beads which is folded back on itself via ionic interactions [17] (see Figure 2.2). This compact form opens up into an extended form upon increasing the ionic strength [89] above 0.75 M. Upon addition of denaturant [66, 2] or increase of pH [89], Fn modules unfold, which can be observed by CD spectroscopy, fluorescence anisotropy, fluorescence emission [88] and electron spin resonance [89] techniques. For GdnHCl-induced denaturation of intact Fn, a gradual unfolding transition was observed between 0 M and 1.2 M GdnHCl, which was not observed in fragments of Fn [88]. Thus, intramolecular interactions appear to stabilize part of the secondary structure of Fn modules.

Owing to the complex interplay of Fn extension and unfolding, previous studies using Fn FRET probes were not able to distinguish loss of secondary structure form loss of intramolecular interactions [9, 7, 76].

Compared to these studies, a more than two-fold higher degree of donor labeling was used. This was desirable, since the nanofibrils under scrutiny are much thinner than the diffraction-limited focal spot of the confocal microscope [90]. A higher degree of donor labeling led to an increased fluorescence signal, which allowed greater measurement accuracy and lower photobleaching. This high density of labeling might lead to self-quenching of donor molecules [91] and, consequently, a lower sensitivity towards molecular unfolding. However, the fact that the FRET signal did not change when extending Fn molecules in high ionic strength buffer (see Figure 10.5) leads to the conclusion that, rather, the sensitivity towards molecular unfolding was pronounced compared to extension of the Fn molecule.

FRET Calibration in Solution Displays Concentration Dependence

Preliminary calibration efforts involved FRET measurements of labeled Fn in solution. These measurements showed a reduction of FRET signal upon denaturation as observed for Fn immobilized in a polyacrylamide gel. However, the absolute values appeared to depend on the protein concentration used. This can be attributed to the sampling procedure of the confocal microscope, which might lead to clipping of low light signal levels and thus an altered FRET ratio, as shown in Figure A.3.

10.2 Determination of Ratio between FRET Labeled and Unlabeled Fn

In order to rule out intermolecular FRET, the concentration of labeled Fn was chosen after an analysis of the FRET signal derived from nanofibrils containing increasing amounts of labeled Fn. Based on this assay, a ratio of 1.25 % labeled Fn and 98.75 % unlabeled Fn was subsequently used.

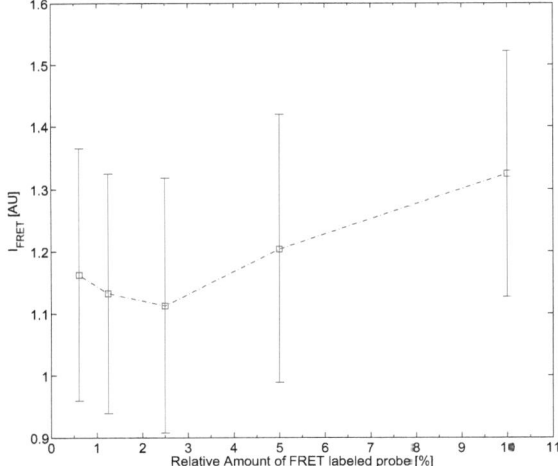

Figure 10.6: Optimal Ratio Between FRET Probe and Unlabeled Fn. The FRET index was determined for nanofibrils containing increasing amounts of FRET probe. Above a probe content of 5 %, the FRET index increases due to intermolecular FRET. The error bars display the standard deviation over all sampled pixels.

10.3 FRET of Fn Surface Films

The formation of a surface film of Fn precedes nanofibril formation. Since protein adsorption to the air-buffer interface is likely to induce molecular unfolding, the degree of Fn unfolding in a surface film was assessed, as shown in Figure 10.7.

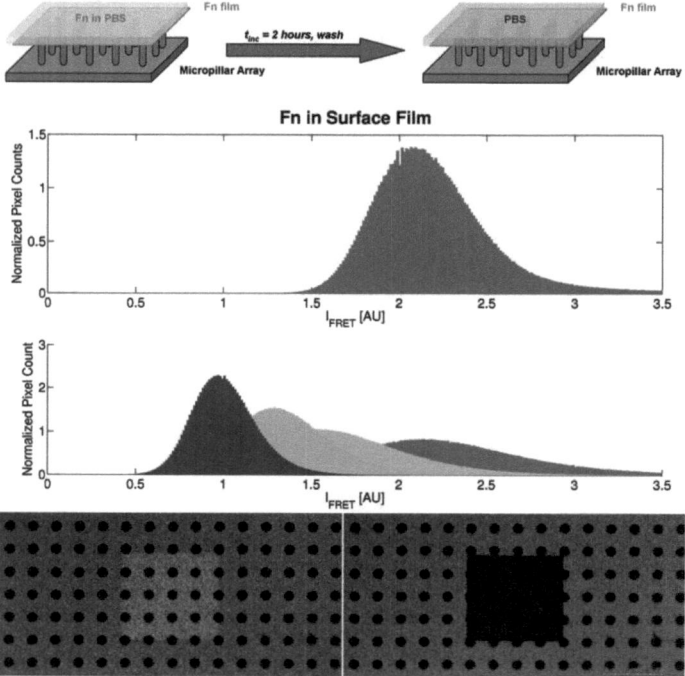

Figure 10.7: FRET Analysis of Fn Surface Films. Top: A film of Fn forms after incubation of Fn solution on top of a superhydrophobic micropillar array. The Fn film can then be analyzed using an upright Confocal Laser Scanning Microscope (CLSM) setup. Middle: During surface accumulation of Fn, no detectable unfolding of Fn takes place, as derived from comparison with the calibration histogram. The bottom histogram shows the calibration curves for increasing GdnHCl concentration. Red: 0 M, gold: 1 M, cyan: 2 M, blue: 4 M. Bottom: Fn molecules within the surface layer are immobile on a time scale of thirty minutes as seen by the persisting pattern of acceptor photobleaching in the center portion of the image. Additionally, the acceptor photobleaching led to donor dequenching, an indication of FRET.

10.3.1 Discussion

Result in Literature Context

No unfolding of Fn within surface films was detected using the FRET method described above. In contrast to this, the fact that Fn molecules undergo structural alterations upon adsorption has been demonstrated for various interfaces [92, 93, 94, 95, 76, 96, 97]. While this has been shown to effect cell response [77], it is unclear which Fn modules unfold and to which extent they do so.

However, the only study on Fn molecular structure at the air-buffer interface showed high-resolution SEM images of disc-shaped Fn molecules, which resemble the structures observed in solution [72], as shown in Figure 10.8. In summary, this indicates that Fn molecules do not unfold upon accumulation at the air-buffer interface. While this can be concluded for the modules immediately neighboring the FRET acceptor labeling sites, the greater part of the Fn molecule is not covered in this investigation (see Figure 10.1). It is well imaginable that structural changes undetected by FRET occur which precede nanofibril formation.

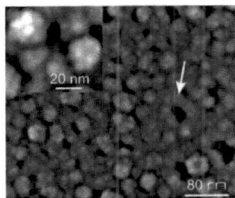

Figure 10.8: High Resolution SEM Image of Fn Molecules at the Air-Buffer Interface. After fixation, critical point drying and coating with 2 nm Chromium, the structure of Fn molecules at the air-buffer interface appears disc-shaped. The inset shows a magnification of the area marked by the white arrow. Taken from [72].

Experiments on porcine Fn using High Performance Liquid Chromatography (HPLC) indicate that the N-terminal 14 kDa heparin-binding fragment as well as the gelatin-binding fragment have the highest degree of hydrophobicity among all Fn fragments [98]. Both fragments have been shown to play a central role in Fn, potentially forming the interaction site between Fn molecules [17] (see Figure 2.2). Since the FRET acceptor fluorophores are restricted to the large central cell binding domain of Fn, neither of those two fragments is probed by the FRET method used in this work.

Existing Models of Fn Fibrillogenesis *in Vitro*

Baneyx and Vogel [26] demonstrated the self-assembly of Fn networks underneath a lipid monolayer followed by gradual expansion of the film [26]. In this work, the authors suggested the insertion of the above mentioned N-terminal fragments into the air-buffer interface, while the remaining fragments, such as the large cell binding fragment, which also carries FRET acceptors, faces towards the subphase.

This model of Fn conformational change upon surface adsorption could explain Fn wetting-induced fibrillogenesis of Fn by exposure of self-binding sites in the N-terminal domains of Fn. However, given the unspecific nature of the labeling and the fact that the N-terminal domains of Fn are not probed using the FRET approach, the current results neither provide proof nor disproof of this model.

10.4 Fn Nanofibrils on PU

From the FRET analysis of Fn films, it was concluded that Fn molecules are not in an unfolded state when assembled at the air-buffer interface. The question was whether the surface forces and sample dehydration that result from the nanofibril fabrication process would unfold Fn molecules during fibrillogenesis.

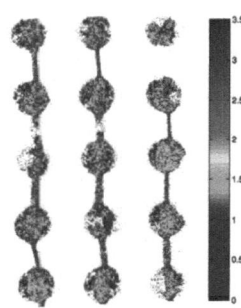

Figure 10.9: FRET Analysis of Fn Nanofibrils on PU Micropillars. For the subsequent analysis of Fn unfolding in nanofibrils, a region of interest was defined, which contained only the nanofibrils, but not the pillar tops. The color map corresponds to the calibration images, i.e. the color values for increasing denaturant concentration (0 M, 1 M, 2 M and 4 M GdnHCl) are red, gold, cyan and blue, respectively. One pillar diameter corresponds to 10 μm.

10.4.1 Discussion

Fn Molecular Unfolding During Nanofibril Formation

During nanofibril formation, molecular unfolding of Fn appears. Given the fact that Fn accumulation at the air-buffer interface does not result in a change in FRET signal, this can either be attributed to surface forces acting during the fabrication process or denaturation of protein upon the inherent dehydration of the nanofibrils. A distinction between both causes is currently not possible.

Molecular Unfolding Is not Caused by Addition of Surfactant

Surface wetting after nanofibril formation proceeds spontaneously on PU micropillars. This demonstrates that the increase in molecular unfolding can not be attributed to the presence of surfactant (MOWIOL-488) during the wetting process. In the experiments described in the following section, an addition of MOWIOL-488 was necessary on PDMS micropillar arrays to allow wetting of the superhydrophobic surfaces.

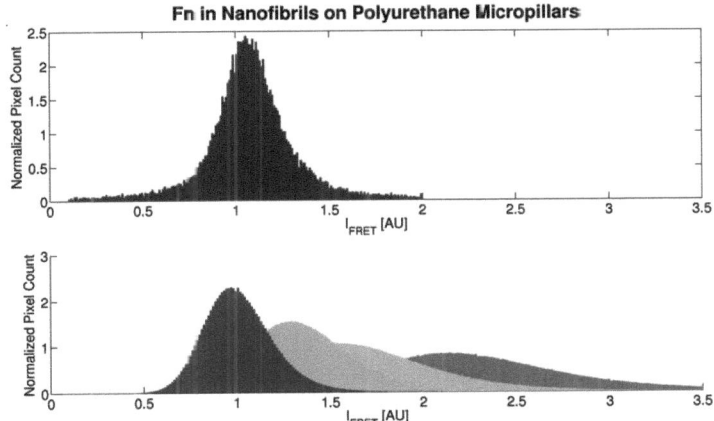

Figure 10.10: FRET Analysis of Fn Nanofibrils on Polyurethane Micropillar Arrays. The degree of Fn molecular unfolding increases drastically during nanofibril formation as compared to the surface film of Fn shown in Figure 10.7. The bottom histogram shows the calibration curves for increasing GdnHCl concentration. Red: 0 M, gold: 1 M, cyan: 2 M, blue: 4 M.

Implications for Structural and Mechanical Properties

It should be clear from the arguments laid out in Section 10.1.4 that the term unfolding represents a rather ill defined process for the Fn molecule. The only conclusion which can be derived with certainty is that the modules around the acceptor sites ($FnIII_7$ and $FnIII_{15}$) unfold during the process. The unfolding of other modules, for example the main cell-binding modules $FnIII_{9-10}$, is suggested by previously published work on Fn denaturation [66, 88, 98], but can not be concluded from the FRET measurements described above. However, since these FnIII modules were shown to be mechanically less stable than the ones probed using the FRET approach [33] (see Figure 2.4, it is highly likely that this hypothesis will be proven right.

10.5 Fn Nanofibrils on Stretchable Substrates

From the FRET analysis of Fn nanofibrils on PU microarrays, it could be concluded that Fn molecules are already in an unfolded state. The question was whether the folded state could be altered by stretching or relaxing Fn nanofibrils. To this end, a PDMS micropillar array was cast and cured on top of a stretchable silicone sheet. After fabrication of Fn nanofibrils and surfactant-induced wetting, this array was stretched up to 300 % elongation (Figure 10.11):

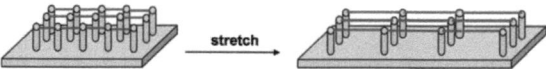

Figure 10.11: Fn Nanofibril Stretching. After wetting of the surface in PBS, the nanofibrils were gradually stretched to a total length of 300 % over a time span of one hour.

FRET signals were recorded for the relaxed, stretched and ruptured state of nanofibrils, as shown in Figures 10.12 and 10.13.

10.5.1 Discussion

Fn Unfolding Increases Upon Stretching

From the results described in the previous sections, it can be derived that Fn adsorbed to the air-buffer interface unfolds drastically when forming nanofibrils. Upon stretching, Fn molecules within nanofibrils further unfold. This is shown in Figure 10.13, where the mean FRET index is shifted to lower values upon stretching.

Only limited Fn Refolding After Nanofibril Rupture

While mechanical relaxation of nanofibrils after rupture was observed (see Figure 10.12), the complete molecular refolding of Fn constituting those fibrils could not be detected. Rather, the Fn molecules in nanofibrils refolded to a state corresponding to a GdnHCl concentration between 2 M and 4 M.

This result is in line with observations made by Smith *et al.* [10], who found that Fn within cell-derived and artificial fibrils does not completely refold to the solution structure. However, in contrast to the findings presented in their work, refolding of Fn to a degree corresponding to the solution structure at 1 M GdnHCl was not observed in this work. This could be due to irreversible unfolding of individual Fn molecules. Possible causes for this are the forces acting during nanofibril formation or the inherent dehydration of nanofibrils. Indeed, the FRET signal derived for nanofibrils corresponds to a significantly higher degree of unfolding than the cell-derived nanofibrils investigated by Smith *et al.* In their experiments, the degree of

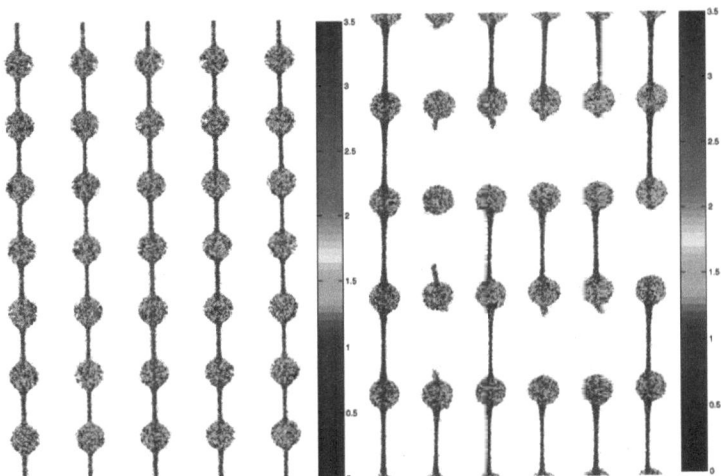

Figure 10.12: FRET Analysis of Relaxed (Left), Stretched and Ruptured (Right) Fn Nanofibrils. After extension of stretchable nanofibrils to 200 % of their original length, Fn nanofibrils begin to rupture. Upon rupture, the nanofibrils contract to one third of their original length, but Fn within those ruptured nanofibrils does not re-fold on a time scale of four hours. One pillar diameter corresponds to 10 µm.

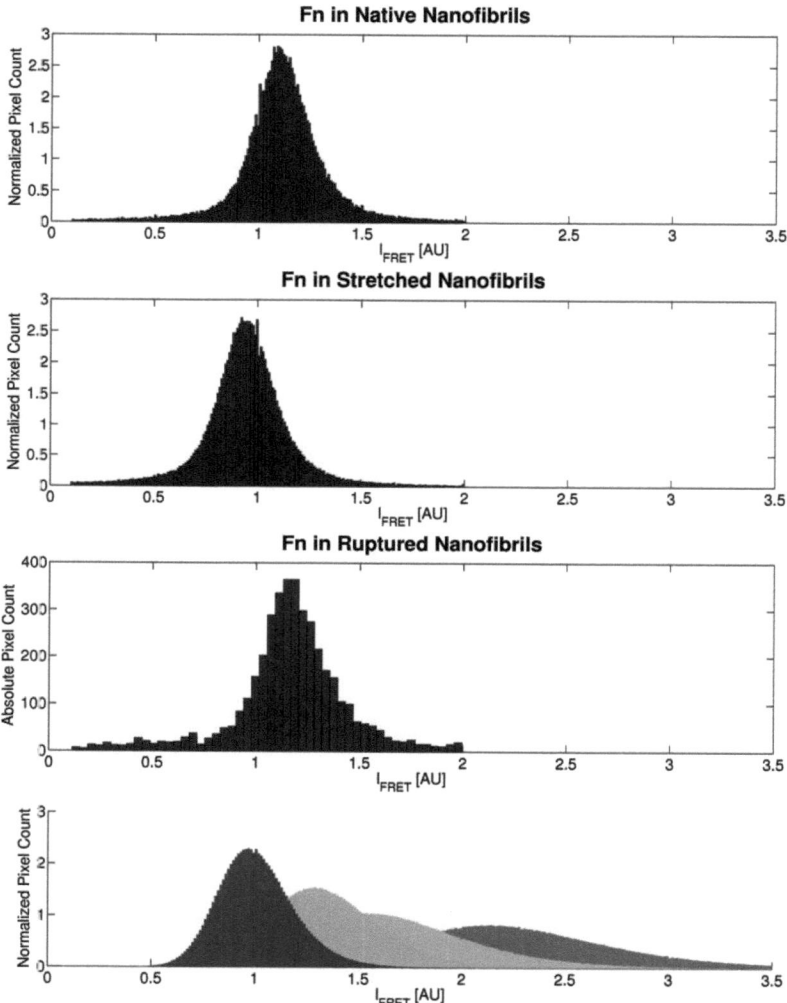

Figure 10.13: FRET Analysis of Fn Nanofibrils on Stretchable PDMS Microarrays. After extension of stretchable nanofibrils to 200 % of their original length, a slight increase in the degree of unfolding of Fn molecules is detectable. When analyzing the ruptured and relaxed nanofibrils only, no re-folding can be observed compared to native nanofibrils. The bottom histogram shows the calibration curves for increasing denaturant concentration. Red: 0 M GdnHCl, gold: 1 M GdnHCl, cyan: 2 M GdnHCl, blue: 4 M GdnHCl.

Fn unfolding observed in cell-derived fibrils corresponds to that of Fn unfolded in 1 M GdnHCl. In contrast, the FRET analysis described here shows that Fn in nanofibrils is unfolded to an extent similar to that found in 4 M GdnHCl. This promotes the hypothesis that "overstretching" of Fn molecules occurs during fibrillogenesis, which leaves the molecules within Fn nanofibrils in an irreversibly unfolded state.

Chapter

11

Mechanical Properties of Fn Nanofibrils

The mechanical properties of cell environments can strongly influence cell behavior such as proliferation and differentiation. However, given the complexity of cell-derived ECM, the mechanical properties of single Fn fibrils have not been analyzed. The free-hanging Fn nanofibrils were thus mechanically probed in this work.

From the experiments on stretchable substrates described in Section 10.5, the elongation at break could be determined for the produced Fn nanofibrils.

In order to derive the material properties of single Fn nanofibrils, three-point bend tests were performed on an AFM setup. Combining force-deflection data from AFM measurements with nanofibril diameters derived in subsequent SEM analysis, the effective Young's Modulus of the produced Fn nanofibrils could be determined.

11.1 Extensibility of Fn Nanofibrils

The elastic properties of cell-derived and artificial Fn fibrils have been described phenomenologically [99, 11]. The hypothesis of this work was that stretching or relaxation of Fn nanofibrils can alter the molecular folding state of Fn. A first important parameter in our mechanical study was the elongation at break and the relaxation behavior of Fn nanofibrils. From the results shown in Figure 11.3, it can be concluded that the extensibility of the produced nanofibrils is rather low, with more than half of the fibrils breaking before extension to 200 % of their original length. The relaxation of Fn could also be observed after nanofibril production on pre-stretched substrates, indicating a pre-strain of 300 %.

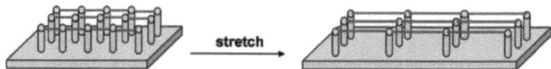

Figure 11.1: Fn Nanofibril Stretching. After wetting of the surface in PBS, the nanofibrils were gradually stretched to a total length of 300 % over a time span of one hour.

Extensibility of Fn Nanofibrils 81

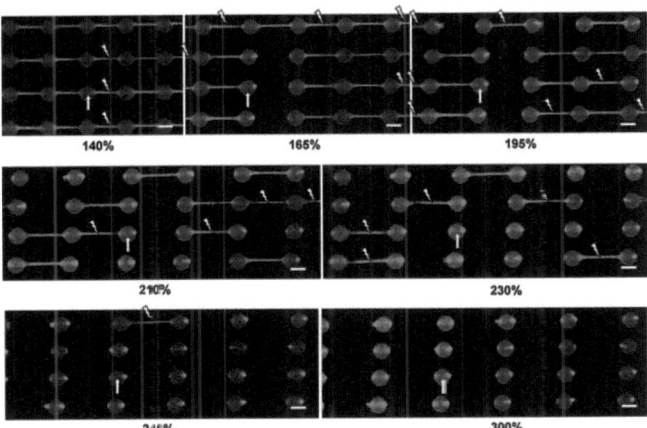

Figure 11.2: Fn Nanofibril Stretching and Breakage. At an extension above 200 %, most of the nanofibrils break. As a guide to the reader, one micropillar is marked with a white arrow. Nanofibrils that are about to undergo rupture are marked by a white lightning bolt in each image. Scale bar: 10 µm

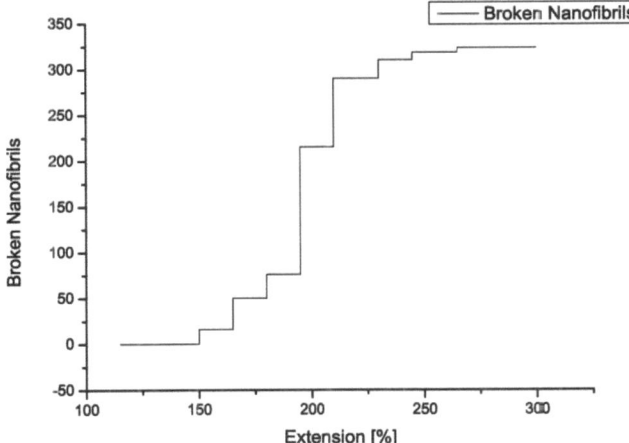

Figure 11.3: Fn Nanofibril Breakage. One field of view comprising 324 nanofibrils was observed during stretching from 115 % to 300 % extension. Between 195 % and 225 % strain, most nanofibrils rupture.

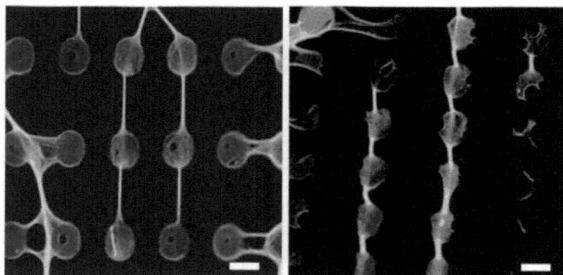

Figure 11.4: Fn Nanofibril Relaxation. Fn nanofibrils were produced on a pre-stretched silicon micropillar array and the strain was released after wetting of the surface. A relaxation to 30 % of their original length was observable before nanofibril buckling, which is in line with the observed relaxation of ruptured nanofbrils. Scale bar: 10 μm.

11.1.1 Discussion

Fn Rupture in the Context of FRET experiments

Compared to the fully relaxed state, the extensibility of cell-derived Fn fibrils was found to be above 400 % [99]. A detailed analysis of micron-sized fibrils produced *in vitro* concluded an elongation at break between 600 % and 1200 % [11].

The nanofibrils produced in this work show an elongation at break of 200 % (Figure 11.3) compared to their original length after fibril formation. Taking into account a relaxation of nanofibrils to 35 % of their original length (Figure 11.4) upon strain release, the overall elongation at break is $\frac{200}{0.35} \approx 600\%$. This is comparable to the values described in literature [99, 11].

Based on the findings described above, a more than three-fold pre-strain within the produced nanofibrils has to be assumed. This is in line with previous FRET-based observations that Fn molecules within nanofibrils unfold during fibrillogenesis [10, 11]. However, in the work described by Little *et al.* [11], the FRET signal at comparable absolute strain corresponds to a lower degree of Fn unfolding than described in Section 10.5. This indicates that differences in the molecular structure of Fn exist between the Fn fibrils produced in the work by Little *et al.* and the nanofibrils reported herein.

Consequences for AFM Experiments

Fn nanofibrils appear to be pre-stressed. This leads to a higher effective Young's modulus measured by our experimental approach. Additionally, the findings from FRET measurements on nanofibrils formed on PU micropillar arrays suggest that a significant portion of the protein within the nanofibrils is unfolded. Thus, the transfer of these results from nanofibrils formed *in vitro* to Fn fibrils found *in vivo* is only possible when taking these facts into consideration.

11.2 AFM and SEM Analysis of Fn Nanofibrils

11.2.1 Experimental Approach

Fn nanofibrils spanned the distance between two adjacent pillars, which allowed the collection of force-deflection curves of Fn nanofibrils in an AFM setup (see Figure 11.5).

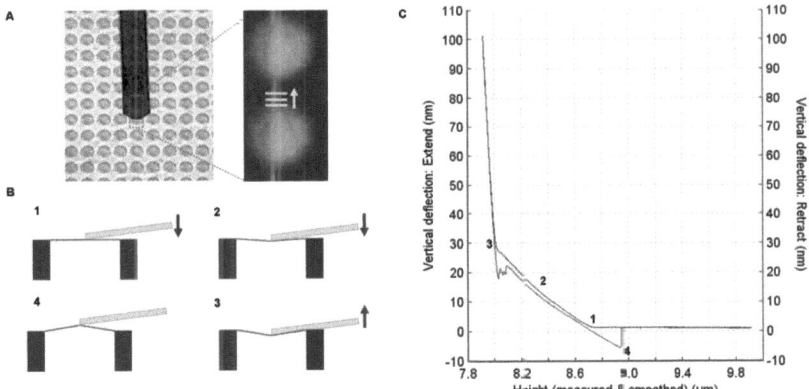

Figure 11.5: Experimental AFM Approach. A: A tipless cantilever probed several positions along a Fn nanofibril, as indicated by the yellow arrow in the fluorescence image. The horizontal lines (yellow) indicate the contact line of the AFM cantilever at different positions along the nanofibril. B: Scheme of one force curve acquisition. C: AFM Cantilever deflection versus piezo height for a full extend (blue) and retract (red) cycle. The numbers correspond to the situations depicted in (B). 1: Contact point, where the cantilever first touches the nanofibril; 2: Bending of the nanofibril; 3: Second contact point where the cantilever touches the stiff micropillar; 4: Detachment point, where the nanofibril adhesion is released again.

Force (F) and deflection (z) values were derived as follows:

$$z = A - D = A - SV, \tag{11.1}$$
$$F = Dk \qquad = SVk, \tag{11.2}$$
$$\tag{11.3}$$

where A is the AFM piezo height signal, D is the cantilever deflection, S is the sensitivity of the cantilever, V is the photodiode voltage signal, and k is the spring constant of the cantilever.

Seen from the perspective of technical mechanics, Fn nanofibrils are nanosized beams, which are fixed at both ends. Application of beam bending theory is possible using equation 11.4 [57], which assumes an isotropic rod with negligible shear modulus:

$$dz = \frac{dF x^3 (l-x)^3}{3l^3 E I_y} \tag{11.4}$$

$$I_y = \frac{\pi d^4}{64} \tag{11.5}$$

$$\frac{dF}{dz} = \frac{3l^3 E I_y}{x^3 (l-x)^3}, \tag{11.6}$$

where z is the beam deflection parallel to the force, F, at position x, and I_y is the area moment of inertia of a cylindrical beam.

Figure 11.6 graphically illustrates the characteristics of equation 11.6: The closer the loading point lies to the fixed end of the beam, the more force is needed to deflect the beam. Thus, the slope in the force-deflection curve, dF/dz, will vary with position, x, according to equation 11.6.

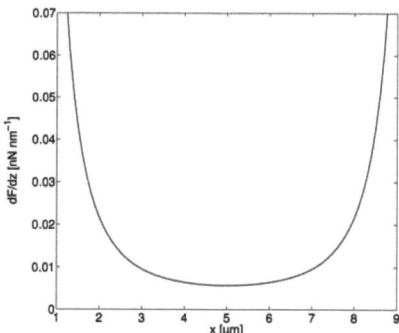

Figure 11.6: Beam Deflection as a Function of Loading Position. The graph shows the expected beam deflection as a function of loading position, according to equation 11.4, with d = 150 nm, E = 1 GPa, l = 10 μm.

11.2.2 AFM Analysis

Force curves were collected along Fn nanofibrils (see Figure 11.5). Each force curve was corrected for offset and tilt using a linear fit to the baseline region of the extension part of the curve. After conversion of AFM photodiode voltage into force acting on the cantilever (see equation 11.3), the contact point between cantilever and nanofibril was determined semi-automatically and a line was fitted to the first 100 nm of the force-deflection signal. The slope of the fit, dF/dz, was plotted as a function of force scan position along the nanofibril. To derive the parameters x_0 and E, equation 11.6 [57, 100], was fitted to the plot of dF/dz versus x, as shown in Figure 11.7.

11.2.3 SEM Analysis and Resulting Bending Moduli

Nanofibril diameters (Figure 11.8) and effective bending moduli were determined for a total of 35 nanofibrils over three independent experiments. The results are summarized in Figure 11.9.

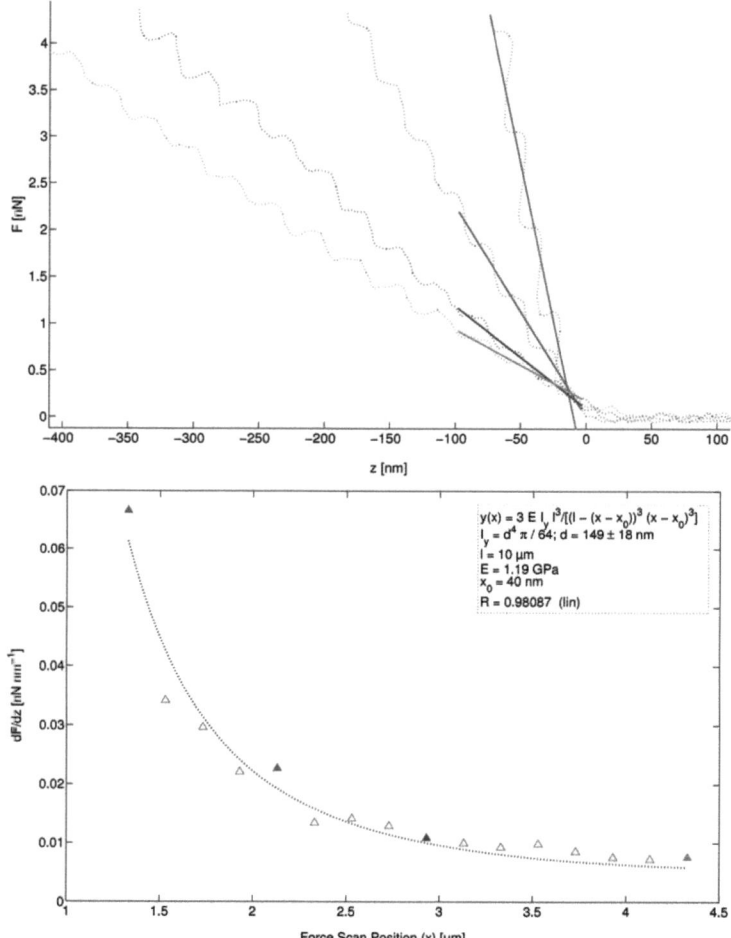

Figure 11.7: Single Nanofibril AFM Analysis. The upper graph shows force-deflection curves derived from AFM experiments as a function of loading position. As the distance from the micropillar increases, the slope of the force-deflection curves decreases. The first 100 nm of each curve were fitted to a line and the dF/dz value was used for the subsequent fit routine. The dF/dz values derived from the depicted curves are shown in their respective colors in the graph below.

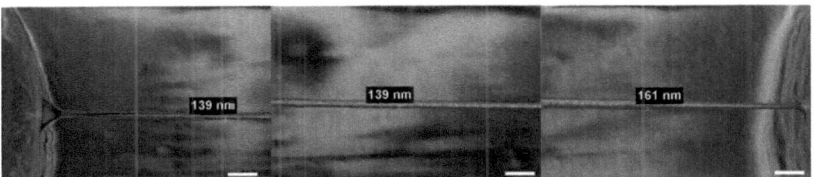

Figure 11.8: Fn Nanofibril After AFM Experiment. To illustrate the variability of Fn nanofibril diameter, the montage shows SEM images along a single nanofibril. Scale bar: 500 nm.

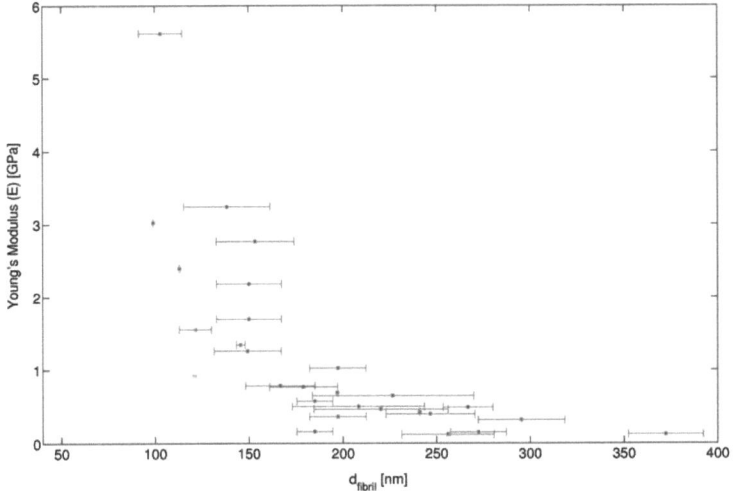

Figure 11.9: Effective Bending Modulus as a Function of Fn Nanofibril Diameter. Error bars correspond to the standard deviation of diameter measurements at three random positions along each nanofibril. The variations in nanofibril diameter along a single fibril are the main source of error in the experiment, as detailed in the discussion.

11.3 Discussion

11.3.1 Comparison of Results to Related Biopolymers

Table 11.1: Stiffness (Young's modulus, E) and breaking strain (extensibility, ϵ_{max}) of various protein fibers. Adapted from Carlisle et al. [101].

Material	E [MPa]	ϵ_{max} [%]
Fibronectin	9*	3
Electrospun fibrinogen fibers	7	1.3
Fibrin fibers		
crosslinked	15	
uncrosslinked	2	2.3
partially crosslinked		3.3
Electrospun COL I fibers (crosslinked)	0.07-0.26	
Tendon collagen (mammalian tendon)	160-7500	0.12
Elastin (bovine ligament)	1	1.5
Resilin		
dragonfly tendon	1-2	1.9
cloned		3.9
Matrix-free Intermediate filament		
mammalian	6-300	1.6
hagfish		2.2
Fibrillin	0.2-100	< 1.9
Myofibrils (sarcomere), titin (connectin)	1	2
Actin	1800-2500	≤ 0.15
Microtubules	1000-1500	≤ 0.20

*: Lorenz Rognoni, personal communication.

The reported mechanical properties of various protein fibrils are summarized in Table 11.1. A distinction should be made between tissue and cell-derived protein fibrils and their counterparts derived *in vitro*. Generally, protein fibrils *in vivo* appear to be softer and more elastic.

Most data on the mechanical properties of ECM nanofibrils such as COL I or fibrin stems from measurements on electrospun nanofibrils, a process which involves drying of protein from an organic solvent and likely leads to protein denaturation.

Compared to other highly extensible ECM protein fibrils, Fn nanofibrils display a very high effective Young's modulus between 0.1 and 6 GPa (Figure 11.9). Recently, values of $E \leq 10$ kPa were gained from nanoindenation experiments on cell-derived Fn fibrils (Lorenz Rognoni, personal communication). Compared to this value, the nanofibrils investigated in this work show an effective Young's Modulus which is four orders of magnitude larger. This could be explained by:

- the presence of pre-stress in Fn nanofibrils,
- changes in the molecular packing of the proteins forming the fibrils, or
- the presence of other ECM components in cell-derived fibrils.

11.3.2 Sources of Error

As seen in Equation 11.6, the largest source of error stems from the determination of nanofibril diameter, since it is present in the fourth power. This is an inherent source of error, since the diameter of nanofibrils varies *per se*, as shown in Figure 11.8.

A second source of error is the data acquisition and subsequent analysis by a least squares fit. Since dF/dz decreases with the third exponent of the distance from the pillar rim (Equation 11.6), there is a trade-off between the accessible position, x, along the fibril, and the accuracy with which it can be probed: The nearer the measurement lies to the pillar rim, the stiffer the AFM cantilever has to be. This sacrifices the possibility to resolve small forces when applying a load at the center of the nanofibril.

Another paramter is the pre-stress under which the nanofibrils find themselves even without application of an external force, which could not be determined experimentally.

Additionally, the model used for the fit procedure assumes an isotropic material with circular and homogeneous cross section. This is, of course, only an approximation to the real situation. Taking these simplifications into account, the AFM approach described here allows an estimation of the effective Young's Modulus of Fn nanofibrils.

Part IV

Conclusions and Outlook

Chapter

12

Protein Structure at Interfaces and Within Nanofibrils

12.1 Conclusions

From the results presented in Section 8.1, it can be concluded that all tested ECM proteins accumulate at the air-buffer interface on a timescale of less than one minute. For the case of Fn, structural changes upon accumulation at the air-buffer interface were not detectable using a FRET-based assay, as shown in Section 10.3. This is in line with prior observations of Fn molecules within identical surface films in high-resolution SEM by Ulmer [85].

Fn molecules within nanofibrils were shown to be unfolded and refolding was not observed using the FRET-based assay. While the results of this thesis indicate that Fn in nanofibrils is in a unfolded form, the FRET approach described does not allow detection of the folding state of relevant cell-binding modules within the Fn molecule.

12.2 Outlook

A method to site-specifically label Fn with fluorescent probes of both donor and acceptor would allow the investigation of specific functional regions of the Fn molecule. This could be achieved by inclusion of Green Fluorescent Protein (GFP) derivatives at defined locations within the Fn gene [102]. Given the large size of the Fn gene and the low protein yield when isolating Fn from cell culture, this approach is both technically demanding and time-consuming.

To investigate the molecular structure of a variety of ECM protein nanofibrils, a label-free method to detect molecular unfolding would be preferable. One method which could fill this gap in the near future is Coherent Anti-Stokes Raman Spectroscopy (CARS). Especially Collagen molecules, which are rich in proline and glycine would yield C–H vibrational contrast. Preliminary experiments in the group

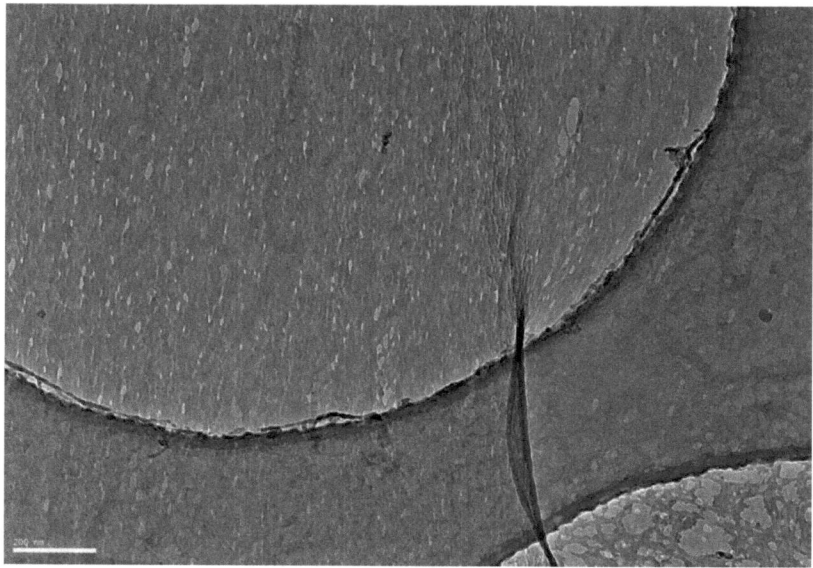

Figure 12.1: Fn Film on TEM Grid. After Langmuir Schaefer transfer of a compressed surface film of Fn molecules, globular structures span the width of a hole in the carbon film. At the center, a fibril forms a bridge between two adjacent holes. Scale bar: 200 nm

of Prof. Andreas Volkmer (University of Stuttgart, personal communication) indicate that COL fibrils show a heterogeneous CARS signal when scanning their long axis. Investigating these properties on COL nanofibrils which are freely suspended could help reveal the nature of this observed contrast.

12.2.1 Use of Protein Films

The fact that Fn molecules appear to be in a native conformation when accumulated at the air-buffer interface could open the route to novel patterning techniques of cell-adhesive substrata.

In preliminary experiments, the transfer of Fn molecules to Transmission Electron Microscopy (TEM) grids was achieved (Figure 12.1). The molecules were visible as globular particles, similar to the particles observed by Ulmer [85]. Using combined fluorescence and high-resolution electron microscopy, this could enable the observation of cell-induced Fn fibrillogenesis in molecular detail.

Chapter 13

AFM Investigation of Fn Fibrillogenesis

13.1 Conclusions

The results described in Sections 8.1 and 10.3 show that it is possible to produce a thin layer of completely folded Fn molecules at the air-buffer interface. This can be used to investigate the process of Fn fibrillogenesis.

13.2 Outlook

New methods to investigate the force-dependence of Fn fibrillogenesis could make use of the film of Fn molecules that was reported in Section 10.3. Preliminary experiments were performed to test the feasibility of the following approach: After touching the film with an adhesive cantilever, single Fn molecules and fibrils can be pulled out of the surface film. This avoids the dehydration of fibrils during stretching and could give valuable insights into the mechanical properties of nascent Fn fibrils (Figures 13.1 and 13.2).

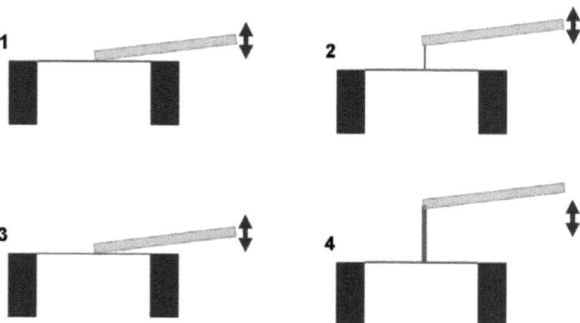

Figure 13.1: Model of AFM-induced Fn Fibrillogenesis from a Surface Film. An AFM tip functionalized with gelatin was repeatedly brought into contact with a surface film of Fn. At each approach-retract cycle, the Fn becomes longer as more molecules attach to the nascent fibrils.

Outlook

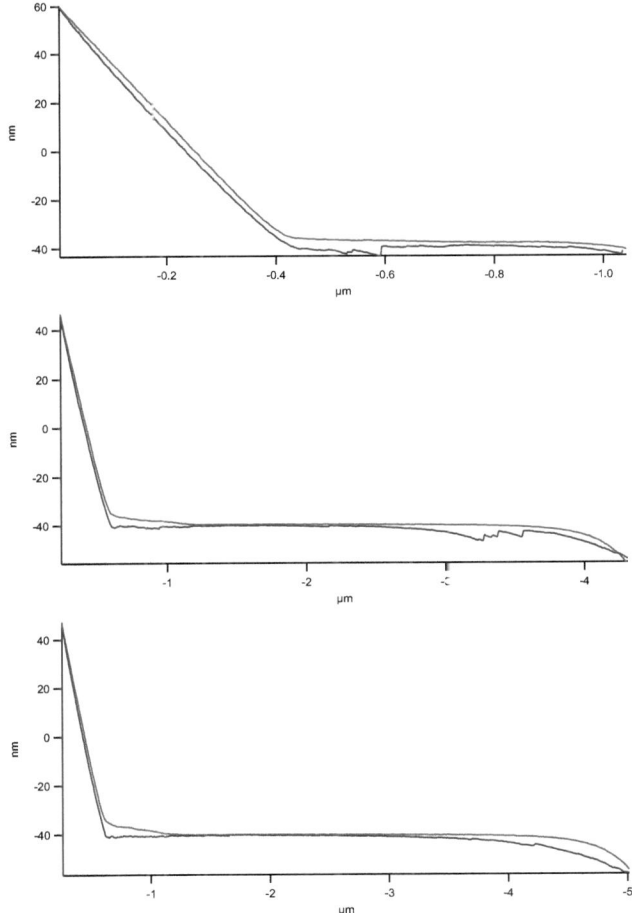

Figure 13.2: AFM-induced Fn Fibrillogenesis from a Surface Film. An AFM tip functionalized with gelatin was repeatedly brought into contact with a surface film of Fn. The resulting force signal shows several unfolding events, during which the Fn molecules gradually and irreversibly unfold. Top: The first force scan shows several unfolding events, which are seen as kinks in the cantilever retract cycle (blue). Middle: Parts that have already been unfolded do not show repeated unfolding, indicating irreversibility of the process. Bottom: The final contour length of above 4 µm indicates that the observation did not only comprise a single Fn molecule, which would have a length of less than 3.5 µm, but rather several molecules which interconnect to form a fibril.

Chapter

14

Production of Polymer Nanofibrils on Micropillar Arrays

14.1 Conclusions

A method for the production of nanofibrils consisting of Fn, COL and LM has been established in this work. As shown in Section 9.3, as well as in work published by other groups [81, 84], the transfer of the nanofibril fabrication method to different biopolymers is possible. This was demonstrated by the formation of actin nanofibrils, described in Section 9.3.4.

Fn nanofibrils rupture at an elongation of 200 % and relax to one third of their original length. This corresponds to a total elongation at break of 600 % relative to the fully relaxed state, in line with findings for Fn fibrils from other groups [99, 11].

While the AFM experiments described in this thesis allow an estimation of the bending modulus of Fn nanofibrils, it is questionable whether the determined absolute values are transferrable to native Fn fibrils. This is mainly due to the fact that Fn within nanofibrils is in a pre-strained and highly unfolded form, which is not observed in native Fn fibrils.

14.2 Outlook

Both cell experimental and materials science experiments are conceivable. On the one hand, the free-hanging nanofibrils of ECM proteins could be used in the investigation of cellular contacts with fibrillar matrices (see Figure 14.1).

On the other hand, since the nanofibril fabrication process is transferable to other polymers, it can be used to study the mechanical properties of a variety of polymer nanofibrils. However, one has to be aware that the values derived from such experiments are not easily transferable to nanofibrils produced using different methods.

Outlook

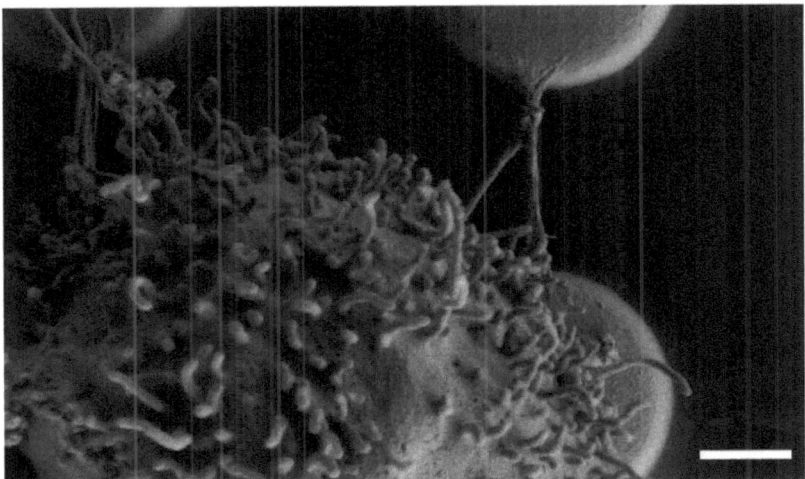

Figure 14.1: KG1a Cell Contacting Fn Nanofibrils. KG1a cells are hematopietic-like stem cells, which are cultured in suspension. After seeding the cells on nanofibril subrates they establish contacts to Fn nanofibrils via tube-like protrusions. Scale bar: 1 µm.

List of Figures

1.1	Extra Cellular Matrix Rigidity of Different Tissues	6
1.2	Model of Cell-ECM Interaction	6
2.1	Fn structure	9
2.2	Proposed Model of Fn Fibrillar Interactions	10
2.3	Fibrils Pulled Out From Fn Solution	12
2.4	Fn Domain Unfolding Hierarchy	13
2.5	Putative Fn Conformations within ECM Fibrils	14
2.6	COL Fibrillogenesis	16
2.7	LM-111 Structure	18
2.8	LM Polimerization	19
3.1	Microarray Fabrication Process	22
3.2	Photoresist Thickness Determination	23
3.3	Effect of Fluorosilane Coating	24
3.4	Demoulding of PDMS Negative from PU	26
4.1	Fabrication of ECM Fibrillar Arrays	27
4.2	Stretching Device Used for Fn Nanofibils	32
5.1	Transfer of ECM Nanofibrils onto PEG	34
6.1	FRET Labeling Scheme of Fn	36
7.1	AFM Sample	43
8.1	Pulling Fibrils from Surface Films of ECM Proteins	48
8.2	Surface Activity of ECM Proteins and BSA	49
8.3	Formation of an Immobile Fn Layer at the Air-Buffer Interface . . .	50
9.1	Silicon Masks	53
9.2	Experiment to Produce Fn Nanofibrils	53
9.3	Fabrication of ECM Fibrillar Arrays	54
9.4	Diameter Control of Fn Nanofibrils	54
9.5	Regular Arrays of Fn Nanofibrils	55
9.6	Model of Nanofibril Formation	56
9.7	Concentration and Pulling Speed dependence of Fn Nanofibrils . . .	57

9.8	Nanofibril Formation Requires Self-Binding Sites	58
9.9	Regular Arrays of Fn Nanofibrils	59
9.10	Western Blot of ECM Protein Preparations	60
9.11	Actin and LM-511 Nanofibrils	61
9.12	Transfer of ECM Nanofibrils onto PEG	62
9.13	Differntial Cell Adhesion on Fn and LM Nanofibrils	63
9.14	Focal Adhesion Staining on Fn and LM Nanofibrils	64
10.1	FRET Labeling Scheme of Fn	65
10.2	CD Analysis of Fn Unfolding	67
10.3	Wavelength Scan of FRET Probe	68
10.4	FRET Analysis of Fn Immobilized in Native PAGE	68
10.5	FRET Analysis of Fn extended in High Ionic Strength Buffer	69
10.6	Optimal Ratio Between FRET Probe and Unlabeled Fn	71
10.7	FRET Analysis of Fn Surface Films	72
10.8	SEM Image of Fn Molecules at the Air-Buffer Interface	73
10.9	FRET Analysis of Fn Nanofibrils on PU Micropillars	74
10.10	FRET Analysis of Fn Nanofibrils on PU Micropillar Arrays	75
10.11	Fn Nanofibril Stretching	76
10.12	FRET Analysis of Relaxed, Stretched and Ruptured Fn Nanofibrils	77
10.13	FRET Analysis of Fn Nanofibrils on Stretchable PDMS Microarrays	78
11.1	Fn Nanofibril Stretching	80
11.2	Fn Nanofibril Stretching and Breakage	81
11.3	Fn Nanofibril Breakage	81
11.4	Fn Nanofibril Relaxation	82
11.5	Experimental AFM Approach	83
11.6	Beam Deflection as a Function of Loading Position	84
11.7	Single Nanofibril AFM Analysis	86
11.8	Fn Nanofibril After AFM Experiment	87
11.9	Effective Bending Modulus as a Function of Fn Nanofibril Diameter	87
12.1	Fn Film on TEM Grid	92
13.1	Model of AFM-induced Fn Fibrillogenesis from a Surface Film	94
13.2	AFM-induced Fn Fibrillogenesis from a Surface Film	95
14.1	KG1a Cell Contacting Fn Nanofibrils	97
A.1	Immunostaining of Fn and COL I Nanofibrils	113
A.2	Laser Stability Test	114
A.3	Calibration of FRET Probe in Solution	115
A.4	Photobleaching of FRET Probe	116

Bibliography

[1] Alberts. *Molecular Biology of the Cell*. Garland Science, 2002. 5

[2] Richard O. Hynes *Fibronectins*. Springer-Verlag New York Inc., 1990. Alexander Rich. 5, 8, 11, 66, 70

[3] Jonathan D Humphries, Adam Byron, and Martin J Humphries. Integrin ligands at a glance. *J Cell Sci*, 119(Pt 19):3901–3903, Oct 2006. 5

[4] Gregory T Christopherson, Hongjun Song, and Hai-Quan Mao. The influence of fiber diameter of electrospun substrates on neural stem cell differentiation and proliferation. *Biomaterials*, 30(4):556–564, Feb 2009. 5

[5] Adam J Engler, Shamik Sen, H. Lee Sweeney, and Dennis E Discher. Matrix elasticity directs stem cell lineage specification. *Cell*, 126(4):677–689, Aug 2006. 5, 6

[6] Viola Vogel and Michael P Sheetz. Cell fate regulation by coupling mechanical cycles to biochemical signaling pathways. *Curr Opin Cell Biol*, 21(1):38–46, Feb 2009. 5, 6, 61

[7] Gretchen Baneyx, Loren Baugh, and Viola Vogel. Fibronectin extension and unfolding within cell matrix fibrils controlled by cytoskeletal tension. *Proc Natl Acad Sci U S A*, 99(8):5139–5143, Apr 2002. 7, 13, 70

[8] Viola Vogel. Mechanotransduction involving multimodular proteins: converting force into biochemical signals. *Annu Rev Biophys Biomol Struct*, 35:459–488, 2006. 7, 9

[9] G. Baneyx, L. Baugh, and V. Vogel. Coexisting conformations of fibronectin in cell culture imaged using fluorescence resonance energy transfer. *Proc Natl Acad Sci U S A*, 98(25):14464–14468, Dec 2001. 7, 12, 13, 65, 70

[10] Michael L Smith, Delphine Gourdon, William C Little, Kristopher E Kubow, R. Andresen Eguiluz, Sheila Luna-Morris, and Viola Vogel. Force-induced

unfolding of fibronectin in the extracellular matrix of living cells. *PLoS Biol*, 5(10):e268, Oct 2007. 7, 12, 13, 14, 65, 66, 76, 82

[11] William C Little, Michael L Smith, Urs Ebneter, and Viola Vogel. Assay to mechanically tune and optically probe fibrillar fibronectin conformations from fully relaxed to breakage. *Matrix Biol*, 27(5):451–461, Jun 2008. 7, 11, 12, 13, 28, 36, 66, 80, 82, 96

[12] William C Little, Ruth Schwartlander, Michael L Smith, Delphine Gourdon, and Viola Vogel. Stretched extracellular matrix proteins turn fouling and are functionally rescued by the chaperones albumin and casein. *Nano Lett*, Sep 2009. 7

[13] M. W. Mosesson and R. A. Umfleet. The cold-insoluble globulin of human plasma. i. purification, primary characterization, and relationship to fibrinogen and other cold-insoluble fraction components. *J Biol Chem*, 245(21):5728–5736, Nov 1970. 8

[14] K. M. Yamada and J. A. Weston. Isolation of a major cell surface glycoprotein from fibroblasts. *Proc Natl Acad Sci U S A*, 71(9):3492–3496, Sep 1974. 8

[15] R. O. Hynes, I. U. Ali, A. T. Destree, V. Mautner, M. E. Perkins, D. R. Senger, D. D. Wagner, and K. K. Smith. A large glycoprotein lost from the surfaces of transformed cells. *Ann N Y Acad Sci*, 312:317–342, Jun 1978. 8

[16] E. L. George, E. N. Georges-Labouesse, R. S. Patel-King, H. Rayburn, and R. O. Hynes. Defects in mesoderm, neural tube and vascular development in mouse embryos lacking fibronectin. *Development*, 119(4):1079–1091, Dec 1993. 8

[17] Ioannis Vakonakis and Iain D Campbell. Extracellular matrix: from atomic resolution to ultrastructure. *Curr Opin Cell Biol*, 19(5):578–583, Oct 2007. 8, 10, 70, 73

[18] Yong Mao and Jean E Schwarzbauer. Stimulatory effects of a three-dimensional microenvironment on cell-mediated fibronectin fibrillogenesis. *J Cell Sci*, 118(Pt 19):4427–4436, Oct 2005. 10, 58

[19] R. O. Hynes and A. T. Destree. Relationships between fibronectin (lets protein) and actin. *Cell*, 15(3):875–886, Nov 1978. 10

[20] Boris Hinz. Masters and servants of the force: the role of matrix adhesions in myofibroblast force perception and transmission. *Eur J Cell Biol*, 85(3-4):175–181, Apr 2006. 10

[21] Simon Jungbauer, Huajian Gao, Joachim P Spatz, and Ralf Kemkemer. Two characteristic regimes in frequency-dependent dynamic reorientation of fibroblasts on cyclically stretched substrates. *Biophys J*, 95(7):3470–3478, Oct 2008. 10

[22] I. U. Ali, V. Mautner, R. Lanza, and R. O. Hynes. Restoration of normal morphology, adhesion and cytoskeleton in transformed cells by addition of a transformation-sensitive surface protein. *Cell*, 11(1):115–126, May 1977. 11

[23] I. U. Ali and R. O. Hynes. Effects of cytochalasin b and colchicine on attachment of a major surface protein of fibroblasts. *Biochim Biophys Acta*, 471(1):16–24, Nov 1977. 11

[24] B. Wòjciak-Stothard, M. Denyer, M. Mishra, and R. A. Brown. Adhesion, orientation, and movement of cells cultured on ultrathin fibronectin fibers. *In Vitro Cell Dev Biol Anim*, 33(2):110–117, Feb 1997. 11

[25] Z. Ahmed and R. A. Brown. Adhesion, alignment, and migration of cultured schwann cells on ultrathin fibronectin fibres. *Cell Motil Cytoskeleton*, 42(4):331–343, 1999. 11

[26] G. Baneyx and V. Vogel. Self-assembly of fibronectin into fibrillar networks underneath dipalmitoyl phosphatidylcholine monolayers: role of lipid matrix and tensile forces. *Proc Natl Acad Sci U S A*, 96(22):12518–12523, Oct 1999. 11, 73

[27] Patricia Rico, Jose Carlos Rodriguez Hernandez, David Moratal, George Altankov, Manuel Monleon Pradas, and Manuel Salmeron. Substrate-induced assembly of fibronectin into networks. influence of surface chemistry and effect on osteoblast adhesion. *Tissue Eng Part A*, Apr 2009. 11

[28] Y. Chen, L. Zardi, and D. M. Peters. High-resolution cryo-scanning electron microscopy study of the macromolecular structure of fibronectin fibrils. *Scanning*, 19(5):349–355, Aug 1997. 11, 14

[29] Klára Briknarová, Maria E Akerman, David W Hoyt, Erkki Ruoslahti, and Kathryn R Ely. Anastellin, an fn3 fragment with fibronectin polymerization activity, resembles amyloid fibril precursors. *J Mol Biol*, 332(1):205–215, Sep 2003. 12

[30] Tomoo Ohashi and Harold P Erickson. Domain unfolding plays a role in superfibronectin formation. *J Biol Chem*, 280(47):39143–39151, Nov 2005. 12

[31] Y. Oberdorfer, H. Fuchs, and A. Janshoff. Conformational analysis of native fibronectin by means of force spectroscopy. *Langmuir*, 16(23):9955–9958, 2000. 12

[32] Andres F Oberhauser, Carmelu Badilla-Fernandez, Mariano Carrion-Vazquez, and Julio M Fernandez. The mechanical hierarchies of fibronectin observed with single-molecule afm. *J Mol Biol*, 319(2):433–447, May 2002. 12, 13

[33] David Craig, Mu Gao, Klaus Schulten, and Viola Vogel. Tuning the mechanical stability of fibronectin type iii modules through sequence variations. *Structure*, 12(1):21–30, Jan 2004. 12, 13, 75

[34] Simon Mitternacht, Stefano Luccioli, Alessandro Torcini, Alberto Imparato, and Anders Irbäck. Changing the mechanical unfolding pathway of fniii(10) by tuning the pulling strength. *Biophys J*, 96(2):429–441, Jan 2009. 12

[35] Lewyn Li, Hector Han-Li Huang, Carmen L Badilla, and Julio M Fernandez. Mechanical unfolding intermediates observed by single-molecule force spectroscopy in a fibronectin type iii module. *J Mol Biol*, 345(4):817–826, Jan 2005. 12

[36] Mu Gao, David Craig, Viola Vogel, and Klaus Schulten. Identifying unfolding intermediates of fn-iii(10) by steered molecular dynamics. *J Mol Biol*, 323(5):939–950, Nov 2002. 12

[37] Elaine P S Gee, Donald E Ingber, and Collin M Stultz. Fibronectin unfolding revisited: Modeling cell traction-mediated unfolding of the tenth type-iii repeat. *PLoS ONE*, 3(6):e2373, 2008. 12

[38] T. Ohashi, D. P. Kiehart, and H. P. Erickson. Dynamics and elasticity of the fibronectin matrix in living cell culture visualized by fibronectin-green fluorescent protein. *Proc Natl Acad Sci U S A*, 96(5):2153–2158, Mar 1999. 12, 13

[39] Meher Antia, Gretchen Baneyx, Kristopher E Kubow, and Viola Vogel. Fibronectin in aging extracellular matrix fibrils is progressively unfolded by cells and elicits an enhanced rigidity response. *Faraday Discuss*, 139:229–49; discussion 309–25, 419–20, 2008. 12, 13

[40] M. J. Williams, I. Phan, T. S. Harvey, A. Rostagno, L. I. Gold, and I. D. Campbell. Solution structure of a pair of fibronectin type 1 modules with fibrin binding activity. *J Mol Biol*, 235(4):1302–1311, Jan 1994. 14, 36, 65

[41] A. R. Pickford, J. R. Potts, J. R. Bright, I. Phan, and I. D. Campbell. Solution structure of a type 2 module from fibronectin: implications for the structure and function of the gelatin-binding domain. *Structure*, 5(3):359–370, Mar 1997. 14, 36, 65

[42] D. J. Leahy, I. Aukhil, and H. P. Erickson. 2.0 a crystal structure of a four-domain segment of human fibronectin encompassing the rgd loop and synergy region. *Cell*, 84(1):155–164, Jan 1996. 14, 36, 65

[43] Peters, Chen, Zardi, and Brummel. Conformation of fibronectin fibrils varies: Discrete globular domains of type iii repeats detected. *Microsc Microanal*, 4(4):385–396, Jul 1998. 14

[44] Karl E Kadler, Adele Hill, and Elizabeth G Canty-Laird. Collagen fibrillogenesis: fibronectin, integrins, and minor collagens as organizers and nucleators. *Curr Opin Cell Biol*, 20(5):495–501, Oct 2008. 13, 15

[45] K. E. Kadler, D. F. Holmes, J. A. Trotter, and J. A. Chapman. Collagen fibril formation. *Biochem J*, 316 (Pt 1):1–11, May 1996. 15

[46] Matthew D Shoulders and Ronald T Raines. Collagen structure and stability. *Annu Rev Biochem*, 78:929–958, 2009. 16

[47] Shawn M Sweeney, Joseph P Orgel, Andrzej Fertala, Jon D McAuliffe, Kevin R Turner, Gloria A Di Lullo, Steven Chen, Olga Antipova, Shiamalee Perumal, Leena Ala-Kokko, Antonella Forlino, Wayne A Cabral, Aileen M Barnes, Joan C Marini, and James D San Antonio. Candidate cell and matrix interaction domains on the collagen fibril, the predominant protein of vertebrates. *J Biol Chem*, 283(30):21187–21197, Jul 2008. 15

[48] David A Cisneros, Jens Friedrichs, Anna Taubenberger, Clemens M Franz, and Daniel J Muller. Creating ultrathin nanoscopic collagen matrices for biological and biotechnological applications. *Small*, 3(6):956–963, Jun 2007. 15, 17

[49] David F Holmes and Karl E Kadler. The 10+4 microfibril structure of thin cartilage fibrils. *Proc Natl Acad Sci U S A*, 103(46):17249–17254, Nov 2006. 17

[50] D. F. Holmes, H. K. Graham, J. A. Trotter, and K. E. Kadler. Stem/tem studies of collagen fibril assembly. *Micron*, 32(3):273–285, Apr 2001. 17

[51] August J Heim, Thomas J Koob, and William G Matthews. Low strain nanomechanics of collagen fibrils. *Biomacromolecules*, 8(11):3298–3301, Nov 2007. 17

[52] Catherine P Barnes, Scott A Sell, Eugene D Boland, David G Simpson, and Gary L Bowlin. Nanofiber technology: designing the next generation of tissue engineering scaffolds. *Adv Drug Deliv Rev*, 59(14):1413–1433, Dec 2007. 17

[53] Markus J Buehler. Nature designs tough collagen: explaining the nanostructure of collagen fibrils. *Proc Natl Acad Sci U S A*, 103(33):12285–12290, Aug 2006. 17

[54] Pieter J in 't Veld and Mark J Stevens. Simulation of the mechanical strength of a single collagen molecule. *Biophys J*, 95(1):33–39, Jul 2008. 17

[55] Joost A J van der Rijt, Kees O van der Werf, Martin L Bennink, Pieter J Dijkstra, and Jan Feijen. Micromechanical testing of individual collagen fibrils. *Macromol Biosci*, 6(9):697–702, Sep 2006. 17

[56] Marco P E Wenger, Laurent Bozec, Michael A Horton, and Patrick Mesquida. Mechanical properties of collagen fibrils. *Biophys J*, 93(4):1255–1263, Aug 2007. 17

[57] Lanti Yang, Carel F C FitiÄľ, Kees O van der Werf, Martin L Bennink, Pieter J Dijkstra, and Jan Feijen. Mechanical properties of single electrospun collagen type i fibers. *Biomaterials*, 29(8):955–962, Mar 2008. 17, 44, 84, 85

[58] Susanne Schéele, Alexander Nyström, Madeleine Durbeej, Jan F Talts, Marja Ekblom, and Peter Ekblom. Laminin isoforms in development and disease. *J Mol Med*, 85(8):825–836, Aug 2007. 17, 18

[59] Monique Aumailley, Leena Bruckner-Tuderman, William G Carter, Rainer Deutzmann, David Edgar, Peter Ekblom, Jürgen Engel, Eva Engvall, Erhard Hohenester, Jonathan C R Jones, Hynda K Kleinman, M. Peter Marinkovich, George R Martin, Ulrike Mayer, Guerrino Meneguzzi, Jeffrey H Miner, Kaoru Miyazaki, Manuel Patarroyo, Mats Paulsson, Vito Quaranta, Joshua R Sanes, Takako Sasaki, Kiyotoshi Sekiguchi, Lydia M Sorokin, Jan F Talts, Karl Tryggvason, Jouni Uitto, Ismo Virtanen, Klaus von der Mark, Ulla M Wewer, Yoshihiko Yamada, and Peter D Yurchenco. A simplified laminin nomenclature. *Matrix Biol*, 24(5):326–332, Aug 2005. 17, 18

[60] L. Schuger. Laminins in lung development. *Exp Lung Res*, 23(2):119–129, 1997. 18

[61] Christian Bökel and Nicholas H Brown. Integrins in development: moving on, responding to, and sticking to the extracellular matrix. *Dev Cell*, 3(3):311–321, Sep 2002. 18

[62] Maria V Tsiper and Peter D Yurchenco. Laminin assembles into separate basement membrane and fibrillar matrices in schwann cells. *J Cell Sci*, 115(Pt 5):1005–1015, Mar 2002. 18

[63] P. D. Yurchenco, E. C. Tsilibary, A. S. Charonis, and H. Furthmayr. Laminin polymerization in vitro. evidence for a two-step assembly with domain specificity. *J Biol Chem*, 260(12):7636–7644, Jun 1985. 19

[64] S. Choi, P.J. Yoo, S.J. Baek, T.W. Kim, and H.H. Lee. An ultraviolet-curable mold for sub-100-nm lithography. *J. Am. Chem. Soc.*, 126(25):7744–7745, June 2004. 25

[65] Colin K Choi, Miguel Vicente-Manzanares, Jessica Zareno, Leanna A Whitmore, Alex Mogilner, and Alan Rick Horwitz. Actin and alpha-actinin orchestrate the assembly and maturation of nascent adhesions in a myosin ii motor-independent manner. *Nat Cell Biol*, Aug 2008. 30

[66] S. S. Alexander, G. Colonna, and H. Edelhoch. The structure and stability of human plasma cold-insoluble globulin. *J Biol Chem*, 254(5):1501–1505, Mar 1979. 38, 70, 75

[67] Scott J. McClellan and Elias I. Franses. Effect of concentration and denaturation on adsorption and surface tension of bovine serum albumin. *Colloids and Surfaces B: Biointerfaces*, 28(1):63–75, April 2003. 47

[68] C. J. van Oss, L. L. Moore, R. J. Good, and M. K. Chaudhury. Surface thermodynamic properties and chromatographic and salting-out behavior of iga and other serum proteins. *Journal of Protein Chemistry*, 4(4):245–263, August 1985. 47

[69] Katharine B. Blodgett. Films built by depositing successive monomolecular layers on a solid surface. *Journal of the American Chemical Society*, 57(6):1007–1022, June 1935. 47

[70] Brian C. Tripp, Jules John Magda, and Joseph D. Andrade. Adsorption of globular proteins at the air/water interface as measured via dynamic surface tension: Concentration dependence, mass-transfer considerations, and adsorption kinetics. *Journal of Colloid and Interface Science*, 173(1):16–27, July 1995. 47

[71] I.V. Turko, I.S. Yurkevich, and V.L. Chashchin. Langmuir-blodgett films of immunoglobulin g for immunosensors. *Thin Solid Films*, 205(1):113–116, October 1991. 50

[72] Jens Ulmer, Benjamin Geiger, and Joachim P. Spatz. Force-induced fibronectin fibrillogenesis in vitro. *Soft Matter*. 4(10):1998–2007, 2008. 50, 52, 55, 73

[73] Okubo and Kobayashi. Surface tension of biological polyelectrolyte solutions. *J Colloid Interface Sci*, 205(2):433–442, Sep 1998. 50, 55

[74] Andrey Tronin, Timothy Dubrovsky, Svetlana Dubrovskaya, Giuliano Radicchi, and Claudio Nicolini. Role of protein unfolding in monolayer formation on air-water interface. *Langmuir*, 12(13):3272–3275, 1996. 50

[75] Valérie Lechevalier, Thomas Croguennec, Stéphane Pezennec, Catherine Guérin-Dubiard, Maryvonne Pasco, and Françoise Nau. Ovalbumin, ovotransferrin, lysozyme: three model proteins for structural modifications at the air-water interface. *J Agric Food Chem*, 51(21):6354–6361, Oct 2003. 50

[76] Loren Baugh and Viola Vogel. Structural changes of fibronectin adsorbed to model surfaces probed by fluorescence resonance energy transfer. *J Biomed Mater Res A*, 69(3):525–534, Jun 2004. 50, 70, 73

[77] A. J. García, M. D. Vega, and D. Boettiger. Modulation of cell proliferation and differentiation through substrate-dependent changes in fibronectin conformation. *Mol Biol Cell*, 10(3):785–798, Mar 1999. 50, 73

[78] Salima Patel, Alain F Chaffotte, Batt Amana, Fabrice Goubard, and Emmanuel Pauthe. In vitro denaturation-renaturation of fibronectin. formation of multimers disulfide-linked and shuffling of intramolecular disulfide bonds. *Int J Biochem Cell Biol*, 38(9):1547–1560, 2006. 51

[79] Tomoo Ohashi, Anne Augustus, and Harold Erickson. Transient opening of fibronectin type iii (fniii) domains: the interaction of the third fniii domain of fn with anastellin. *Biochemistry*, Mar 2009. 51

[80] Christopher A Lemmon, Christopher S Chen, and Lewis H Romer. Cell traction forces direct fibronectin matrix assembly. *Biophys J*, 96(2):729–738, Jan 2009. 53

[81] Jingjiao Guan and L. James Lee. Generating highly ordered dna nanostrand arrays. *Proc Natl Acad Sci U S A*, 102(51):18321–18325, Dec 2005. 55, 56, 62, 96

[82] Jingjiao Guan, Nicholas Ferrell, Bo Yu, Derek J. Hansford, and L. James Lee. Simultaneous fabrication of hybrid arrays of nanowires and micro/nanoparticles by dewetting on micropillars. *Soft Matter*, 3(11):1369–1371, 2007. 55

[83] J. Guan, B. Yu, and L.ÂăJ. Lee. Forming highly ordered arrays of functionalized polymer nanowires by dewetting on micropillars. *Advanced Materials*, 19(9):1212–1217, 2007. 55, 62

[84] Santosh Pabba, Mehdi M. Yazdanpanah, Brigitte H. Fasciotto Totten, Vladimir V. Dobrokhotov, Jeremy M. Rathfon, Gregory N. Tew, and Robert W. Cohn. Biopolymerization-driven self-assembly of nanofiber airbridges. *Soft Matter*, 5(7):1378–1385, 2009. 58, 62, 96

[85] Jens Ulmer. *Quantitative Measurements of Force Distribution in Single and Multi Cellular Systems*. PhD thesis, Ruprecht Karls Universität Heidelberg, 2005. 61, 91, 92

[86] Peter Kaiser and Joachim Spatz. Differential adhesion of fibroblast and neuroblastoma cells on size- and geometry-controlled nanofibrils of the extracellular matrix. *Soft Matter*, 6:113–119, 2010. 63, 64

[87] Thomas Blättler, Christoph Huwiler, Mirjam Ochsner, Brigitte Städler, Harun Solak, Janos Vörös, and H. Michelle Grandin. Nanopatterns with biological functions. *J Nanosci Nanotechnol*, 6(8):2237–2264, Aug 2006. 63

[88] M. Y. Khan, M. S. Medow, and S. A. Newman. Unfolding transitions of fibronectin and its domains. stabilization and structural alteration of the n-terminal domain by heparin. *Biochem J*, 270(1):33–38, Aug 1990. 66, 70, 75

[89] C. S. Lai, N. M. Tooney, and E. G. Ankel. Structure and flexibility of plasma fibronectin in solution: electron spin resonance spin-label, circular dichroism, and sedimentation studies. *Biochemistry*, 23(26):6393–6397. Dec 1984. 70

[90] James P. Pawley. *Handbook of Biological Confocal Microscopy*. Springer-Verlag New York Inc., 2006. 70

[91] X. Zhuang, T. Ha, H. D. Kim, T. Centner, S. Labeit, and S. Chu. Fluorescence quenching: A tool for single-molecule protein-folding study. *Proc Natl Acad Sci U S A*, 97(26):14241–14244, Dec 2000. 70

[92] C. Narasimhan and C. S. Lai. Conformational changes of plasma fibronectin detected upon adsorption to solid substrates: a spin-label study. *Biochemistry*, 28(12):5041–5046, Jun 1989. 73

[93] D. J. Iuliano, S. S. Saavedra, and G. A. Truskey. Effect of the conformation and orientation of adsorbed fibronectin on endothelial cell spreading and the strength of adhesion. *J Biomed Mater Res*, 27(8):1103–1113, Aug 1993. 73

[94] Benjamin G Keselowsky, David M Collard, and AndrÃĺs J GarcÃŋa. Surface chemistry modulates fibronectin conformation and directs integrin binding and specificity to control cell adhesion. *J Biomed Mater Res A*, 66(2):247–259, Aug 2003. 73

[95] Magnus Bergkvist, Jan Carlsson, and Sven Oscarsson. Surface-dependent conformations of human plasma fibronectin adsorbed to silica, mica, and hydrophobic surfaces, studied with use of atomic force microscopy. *J Biomed Mater Res A*, 64(2):349–356, Feb 2003. 73

[96] Karine Vallieres, Pascale Chevallier, Christian Sarra-Bournet, Stephane Turgeon, and Gaetan Laroche. Afm imaging of immobilized fibronectin: Does the surface conjugation scheme affect the protein orientation/conformation? *Langmuir*, Aug 2007. 73

[97] Laura A Buchanan and Ahmed El-Ghannam. Effect of bioactive glass crystallization on the conformation and bioactivity of adsorbed proteins. *J Biomed Mater Res A*, Jul 2009. 73

[98] A. Hayashi-Nagai, H. Kitagaki-Ogawa, I. Matsumoto, M. Hayashi, and N. Seno. Hydrophobic properties of porcine fibronectin and its functional domains. *J Biochem*, 109(1):83–88, Jan 1991. 73, 75

[99] Tomoo Ohashi, Daniel P Kiehart, and Harold P Erickson. Dual labeling of the fibronectin matrix and actin cytoskeleton with green fluorescent protein variants. *J Cell Sci*, 115(Pt 6):1221–1229, Mar 2002. 80, 82, 96

[100] Lanti Yang, Kees O. van der Werf, Carel F.C. Fitié, Martin L. Bennink, Pieter J. Dijkstra, and Jan Feijen. Mechanical properties of native and cross-linked type i collagen fibrils. *Biophysical Journal*, 94(6):2204–2211, March 2008. 85

[101] Christine R Carlisle, Corentin Coulais, Manoj Namboothiry, David L Carroll, Roy R Hantgan, and Martin Guthold. The mechanical properties of individual, electrospun fibrinogen fibers. *Biomaterials*, 30(6):1205–1213, Feb 2009. 88

[102] Tomoo Ohashi and Harold P Erickson. Revisiting the mystery of fibronectin multimers: The fibronectin matrix is composed of fibronectin dimers cross-linked by non-covalent bonds. *Matrix Biol*, Mar 2009. 91

Appendix

Appendix

A.1 Abbreviations

AFM Atomic Force Microscopy

AOTF Accusto Optical Tunable Filter

APS Ammonium Persulfate

AU Airy Unit

BM Basement Membrane

BSA Bovine Serum Albumin

CARS Coherent Anti-Stokes Raman Spectroscopy

CD Circular Dichroism

CLSM Confocal Laser Scanning Microscope

COL Collagen

DMEM Dulbecco's Modified Eagle's Medium

DMSO Dimethylsulfoxide

DNA Deoxyribonucleic Acid

ECM Extra Cellular Matrix

EDC 1-Ethyl-3-[3-dimethylaminopropyl]carbodiimide hydrochloride

EDTA Ethylenediaminetetraacetate

EHS Engelbreth-Holm-Swarm

FBS Fetal Bovine Serum

Fn Fibronectin

FPLC Fast Protein Liquid Chromatography

FRET Förster Resonance Energy Transfer

GFP Green Fluorescent Protein

GdnHCl Guanidine Hydrochloride

HFF Human Foreskin Fibroblast

HPLC High Performance Liquid Chromatography

LM Laminin

MES 2-(N-morpholino)ethanesulfonic acid

MRE Mean Residue Ellipticity

MRW Mean Residue Weight

MSC Mesenchymal Stem Cell

NSC Neural Stem Cell

NHS N-hydroxysuccinimide

PAGE Polyacrylamide Gelelectrophoresis

PBS Phosphate Buffered Saline

PDMS Poly(dimethylsiloxane)

PEG Polyethyleneglycol

PFA Paraformaldehyde

PNS Peripheral Nerve System

PMSF Phenylmethanesulphonylfluoride

PU Polyurethane

RIE Reactive Ion Etching

SEM Scanning Electron Microscopy

SMD Steered Molecular Dynamics

TC   Tropocollagen

TEM   Transmission Electron Microscopy

TEMED   Tetramethylethylenediamin

UV   Ultraviolet

WLL   White Light Laser

A.2 Immunofluorescence Stain of COL and Fn Nanofibrils

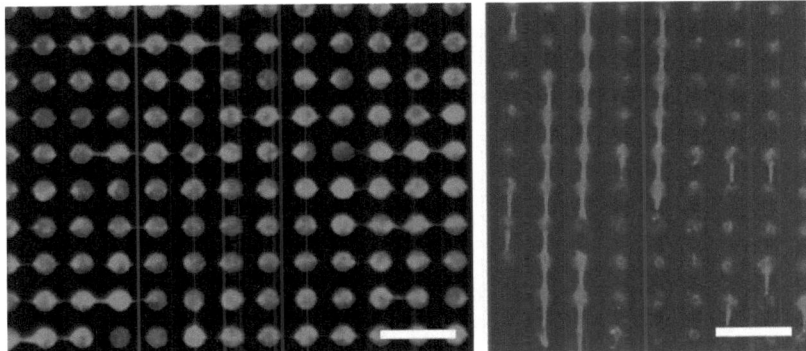

Figure A.1: Immunostaining of Fn and COL I Nanofibrils. Both COL I (green) and Fn (red) Nanofibrils were produced on different areas of a Silicon micropillar array and immunostained. Both images show overlay of the red and green channel. No mutual crosstalk is detectable, indicating that the nanofibrils consist of pure COL I and Fn, respectively.

A.3 Laser Stability Test

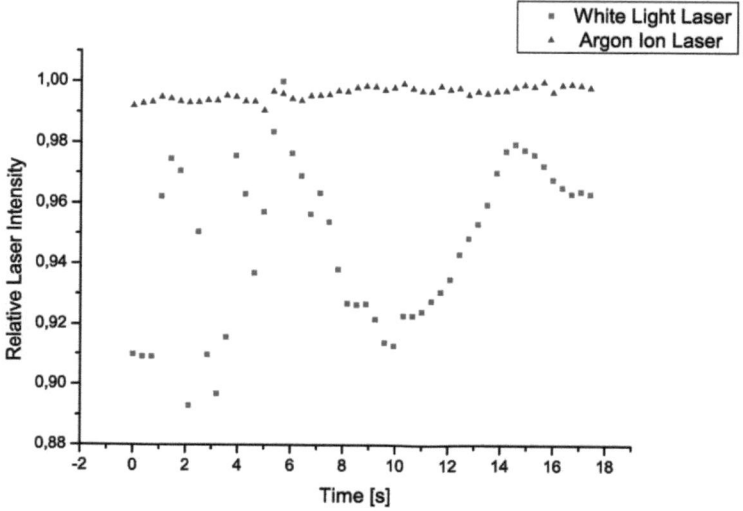

Figure A.2: Laser Stability Test. The stability of two laser sources was tested by focusing on a fluorescent plastic slide and recording the fluorescence signal at constant laser power. The Argon ion laser is a much more stable source than the white light laser and was used in all FRET experiments.

A.4 FRET Calibration in Solution

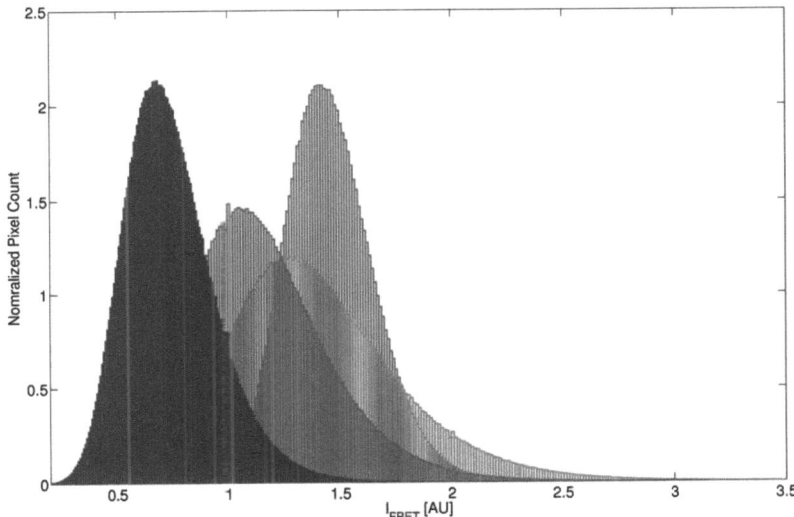

Figure A.3: Calibration of FRET Probe in Solution. The FRET index determined in solution is concentration dependent, as shown by a dilution series of FRET probe in PBS. Red: Stock solution, gold: 0.5 x concentration, green: 0 25 x concentration, blue: 0.125 x concentration. The results indicate that, due to the photomultiplier signal registration at 40 MHz, which is a fixed setting in the CLSM used for this study, low intensity light is clipped in the acceptor channel. This leads to a decrease in I_{FRET} with decreasing fluorescence light levels. The problem could be avoided when calibrating the FRET probe trapped within a polyacrylamide gel.

A.5 Photobleaching during FRET image acquisition

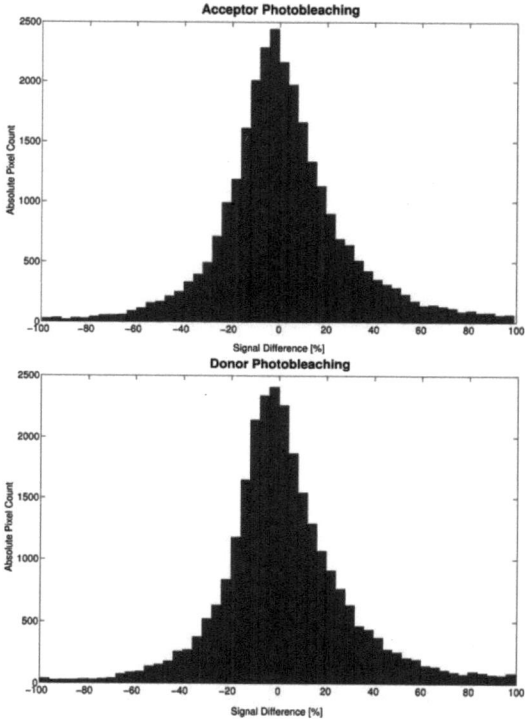

Figure A.4: Photobleaching of FRET Probe. After FRET donor and acceptor image acquisition, a second image was recorded for each channel and the pixel-wise difference in intensity is plotted in the histogram. The mean photobleaching was below 5 % for all images analyzed.

Die VDM Verlagsservicegesellschaft sucht für wissenschaftliche Verlage abgeschlossene und herausragende

Dissertationen, Habilitationen, Diplomarbeiten, Master Theses, Magisterarbeiten usw.

für die kostenlose Publikation als Fachbuch.

Sie verfügen über eine Arbeit, die hohen inhaltlichen und formalen Ansprüchen genügt, und haben Interesse an einer honorarvergüteten Publikation?

Dann senden Sie bitte erste Informationen über sich und Ihre Arbeit per Email an *info@vdm-vsg.de*.

Sie erhalten kurzfristig unser Feedback!

VDM Verlagsservicegesellschaft mbH
Dudweiler Landstr. 99　　　　　　　　Telefon +49 681 3720 174
D - 66123 Saarbrücken　　　　　　　　Fax　　　+49 681 3720 1749
www.vdm-vsg.de

Die VDM Verlagsservicegesellschaft mbH vertritt

Printed by Books on Demand GmbH, Norderstedt / Germany